SWITCHGEAR AND PROTECTION

(For B.Tech./B.E., AMIE, GATE, UPSC Engineering Services)

DR. HAROON ASHFAQ
Asst. Professor
Department of Electrical Engineering,
Faculty of Engineering and Technology,
Jamia Millia Islamia, New Delhi (INDIA)

KHANNA BOOK PUBLISHING CO. (P) LTD.
Publisher of Engineering and Computer Books
4C/4344, Ansari Road, Darya Ganj, New Delhi-110002
Phone : 011-23244447-48 **Mobile:** +91-9910909320
E-mail : contact@khannabooks.com
Website : www.khannabooks.com

Dear Readers,

To prevent the piracy, this book is secured with HIGH SECURITY HOLOGRAM on the front title cover. In case you don't find the hologram on the front cover title, please write us to at contact@khannabooks.com or whatsapp us at +91-99109 09320 and avail special gift voucher for yourself.

Specimen of Hologram on front Cover title:

Moreover, there is a SPECIAL DISCOUNT COUPON for you with EVERY HOLOGRAM.

How to avail this SPECIAL DISCOUNT:

Step 1: Scratch the hologram
Step 2: Under the scratch area, your "coupon code" is available
Step 3: Logon to www.khannabooks.com
Step 4: Use your "coupon code" in the shopping cart and get your copy at a special discount
Step 5: Enjoy your reading!

Thanks,
Khanna Publishing House

Price: ₹ 399.00

Switchgear and Protection
by **Haroon Ashfaq**

ISBN: 978-93-80016-07-8

First Edition: 2009
Reprint: 2024

To view complete list of books, please scan the QR Code:

Published by:
Khanna Book Publishing Co. (P) Ltd.
CIN: U22110DL1998PTC095547

Visit us at: www.khannabooks.com
Write us at: contact@khannabooks.com

Printed in India.

DEDICATION

This work is respectfully dedicated
to the everlasting memory of
my uncle
Abrar Husain

PREFACE

This book is intended to serve as a textbook for the course 'Switchgear and Protection' for B.Tech. / B.E. degree students of Electrical Engineering. It will also serve as a text reference for the students of Diploma in Electrical Engineering. The common topics included in the syllabi of almost all Engineering Institutions in India are covered in this book.

This text is also useful for candidates appearing for AMIE, GATE and UPSC engineering services and other competitive examinations. The practicing engineers would also find this book valuable as a concise reference for basic principles and applications in 'Switchgear and Protection'.

The book is written in a very simple language, so that students may easily grasp the subject. It is ideally suited for self study. It contains sufficient material for one semester undergraduate course. Emphasis has been laid on basic principles. An attempt has been made to make the book as precise as possible.

I hope this book will be useful to students and teachers alike. I will be grateful to readers for their suggestions for the improvement of the book.

HAROON ASHFAQ

CONTENTS

1. **PROTECTION SYSTEMS** **1–17**

1.1 Introduction 1
1.2 Nature and Causes of Faults 2
1.3 Types of Faults 3
1.4 Effects of Faults 4
1.5 Objectives of Power System Protection 5
1.6 Basic Requirements of Protective Relaying 6
1.7 Protective Zones 7
1.8 Classification of Protective Relays 10
1.9 Classification of Protective Schemes 10
1.10 Instrument Transformers 11
1.11 Current Transformer (CT) 12
1.12 Voltage Transformer (VT) or Potential Transformer (PT) 15
Exercises 17

2. **ELECTROMAGNETIC & ELECTRO-THERMAL RELAYS** **18–28**

2.1 Principal Types of Electromagnetic Relays 18
2.2 Attracted Armature Type Relays 18
2.3 Electromagnetic Induction Type Relays 22
2.4 Theory of Induction Relay Torque 25
2.5 Electro-Thermal Relays 27
Exercises 28

3. **STATIC RELAYS** **29–44**

3.1 Introduction 29
3.2 Relay Comparators 30
3.3 Amplitude Comparators 31
3.4 Phase Comparators 35
3.5 Duality between Amplitude and Phase Comparators 41
Exercises 44

4. OVERCURRENT PROTECTION 45–64

4.1 Introduction 45
4.2 Time-Current Characteristics 46
4.3 Setting of Inverse Time Overcurrent Relays 48
4.4 Relay Coordination 51
4.5 Application of Inverse, Time-Delay Overcurrent Relays to a Series of Radial Lines 55
4.6 General Relay Equation 57
4.7 Directional Relays 57
4.8 Reverse Power or Directional Relay 58
4.9 Induction Type Directional Overcurrent Relay 59
4.10 Application of Overcurrent Relays 61
4.11 Application of Directional Overcurrent Relays 61
4.12 Protection of Parallel Feeders 61
4.13 Provision of Directionality to an Overcurrent Relay 62
Exercises 63

5. DISTANCE PROTECTION 65–80

5.1 Reach of Relays 65
5.2 Three-Zone Protection 65
5.3 Ohmic Setting of Distance Relays 67
5.4 Distance Relays 68
5.5 Impedance Relays 69
5.6 Reactance-Type Distance Relay 72
5.7 Admittance Relay (Mho Relay) 75
5.8 Modified Impedance Relay 77
5.9 Ohm Relay 79
Exercises 80

6. TRANSFORMER PROTECTION 81–94

6.1 Differential Protection 81
6.2 Percentage Differential Protection 83
6.3 Transformer Protection 86
6.4 Protection against Magnetizing Inrush Current 87
6.5 Method of Harmonic Current Restraint 88

6.6 Considerations to Protect A Three-Phase Transformer Bank 90
6.7 Desensitizing Scheme for Protection of Transformers 90
6.8 Buchholz Relay 91
6.9 Overfluxing Protection 93
Exercises 94

7. GENERATOR AND MOTOR PROTECTION 95–109

7.1 Introduction 95
7.2 Percentage Differential Protection 96
7.3 Stator Earth Fault-Protection 99
7.4 Stator Interturn Fault Protection 100
7.5 Rotor Earth Fault Protection 101
7.6 Loss of Excitation 102
7.7 Protection against Unbalanced Currents 104
7.8 Overvoltage Protection 105
7.9 Protection against Overheating 105
7.10 Protection against Motoring (Prime Mover Failure) 106
7.11 Unit Generator-Transformer Protection 107
7.12 AC Motor Protection 108
Exercises 109

8. TRANSMISSION LINE AND BUSBAR PROTECTION 110–122

8.1 Pilot Relaying 110
8.2 Pilot-Wire Protection 111
8.3 Power Line Carrier Protection 112
8.4 Phase Comparison Carrier Current Protection 113
8.5 Microwave Pilot 115
8.6 Fibre-Optic Links 115
8.7 Selection of a Communication Channel 115
8.8 Busbar Protection 116
8.9 Differential Protection of a Bus Using Linear Couplers 116

8.10 Differential Protection of a Bus Using High-Impedance Relay 117
8.11 Autoreclosing 118
8.12 Definitions 119
8.13 Single-Shot and Multi-Shot Autoreclosing 120
8.14 Single-Phase (Single-Pole) Autoreclosing 121
8.15 Three-Phase Autoreclosing 121
Exercises 121

9. SYSTEM NEUTRAL GROUNDING 123–138

9.1 Introduction 123
9.2 Undergrounded Neutral System 123
9.3 Arcing Grounds 127
9.4 Advantages of Neutral Grounding 127
9.5 Methods of Neutral Grounding 128
9.6 Solid Grounding 128
9.7 Resistance Grounding 129
9.8 Reactance Grounding 130
9.9 Peterson-Coil Grounding 130
9.10 Grounding Transformer 134
9.11 Choice of Grounding 135
Exercises 137

10. OVERVOLTAGE PROTECTION 139–175

10.1 Introduction 139
10.2 Reflection of waves 140
10.3 Transmission or Refraction of Waves 142
10.4 Open-circuited Line 143
10.5 Short-circuited Line 143
10.6 Line Terminated in Z_0 144
10.7 Junction of Two Dissimilar Lines 144
10.8 Effect of Cable on Surge 150
10.9 Junction of Several Lines 153
10.10 Power System Overvoltages 155
10.11 Internal Overvoltages 155
10.12 External Overvoltages 157

10.13	Mechanism of the Lightning Discharge	157
10.14	Frequency of Lightning Flashovers	158
10.15	Stroke Currents	159
10.16	Direct and Induced Strokes	159
10.17	Waveshapes of Stroke Currents	159
10.18	Overvoltage Protection	160
10.19	Overhead Earth Wire	161
10.20	Rod Gap	163
10.21	Surge Diverters	165
10.22	Expulsion Type Surge Diverter	165
10.23	Valve-Type Surge Diverter	166
10.24	Metal-Oxide Surge Diverter	168
10.25	Location of Surge Diverters	170
10.26	Selection of Surge Diverters	171
10.27	Lightning Performance of Lines	171
10.28	Basic Impulse Insulation Level or Basic Insulation Level (BIL)	171
10.29	Volt-Time Characteristic (Curve)	172
10.30	Insulation Coordination	173
	Exercises	174
11.	**CIRCUIT BREAKING**	**176–194**
11.1	Introduction	176
11.2	Initiation of Arc	177
11.3	Arc Interruption	177
11.4	Arc Extinction Phenomenon	178
11.5	Restriking Voltage and Recovery Voltage	179
11.6	Expression for Restriking Voltage and RRRV	180
11.7	Interruption of Capacitive Current	184
11.8	Current Chopping	186
11.9	Resistance Switching	188
11.10	Interruption of Short Line Fault	193
	Exercises	194

12. CIRCUIT BREAKERS **195–224**

12.1 Classification of Circuit Breakers 195
12.2 Air-Break Circuit Breakers 195
12.3 Air Blast Circuit Breakers (ABCBs) 196
12.4 Oil Circuit Breakers 199
12.5 Sulphur Hexafluoride (SF_6) Circuit Breakers 204
12.6 Vacuum Circuit Breakers 207
12.7 Operating Mechanism 209
12.8 HVDC Circuit Breaking 210
12.9 Ratings of Circuit Breakers 212
12.10 Testing of Circuit Breakers 214
Exercises 223

INDEX 225–233

1

PROTECTION SYSTEMS

1.1 INTRODUCTION

Switchgear is a general term covering switching or interrupting devices and their combination with associated control, instrumentation, metering, protective and regulating devices. It also includes assemblies of these devices with associated interconnections, accessories, and supporting structures.

In power systems, the switchgears are usually located in generating plant switchyards, transmission substations, bulk power substations, and distribution substations.

In a power system the faults should be instantly detected and only the faulty section disconnected from rest of the system in the shortest possible time.

Faults are *detected* automatically by *relays*, and the faulty section *isolated* by circuit breakers.

The combination of relays and circuit breakers is known as the protective systems.

Protection is not needed when the system is operating normally. It is only needed when the system is not operating normally.

A protective system is based on detecting fault conditions by continuously monitoring the power system variables, such as current, voltage, power, frequency, and impedance. It is to be noted that the values of these quantities differ under normal and abnormal conditions. Abnormal values individually or in combination cause relays to operate, energizing tripping circuits of the circuit breakers. Instrument transformers namely, current transformers (CTs) and voltage transformers (VTs), are used to measure currents and voltages. Instrument transformers feed the measured variables to the relay system.

The cost of protection is generally extremely small compared with the cost of equipment protected.

The art of protective relaying is constantly changing and advancing. However, the basic principles of relay operation and application remain the same.

1.2 NATURE AND CAUSES OF FAULTS

Normally, a power system operates under balanced conditions. Under abnormal (that is fault) conditions, the system may become unbalanced. If the insulation of the system fails at any point, or if two or more conductors that normally operate with a potential difference come in contact with each other, a short-circuit, or fault, is said to occur. The contact may be a physical metallic one, or it may occur through an arc.

A fault may occur on a power system due to a number of causes. Some of the common causes have their origin in natural disturbances like lightning, high-speed winds, earthquakes, earth tremors, snow, frost etc. Generators, transformers, and other protective switchgear may fail due to insulation breakdown. There may be accidental faults such as falling of a tree along the lines, vehicles colliding with supporting structures, airplane colliding with the line. Birds may cause faults on overhead lines if their bodies touch one of the phases and the earth wire (or the metallic supporting structure which is at earth potential). Sometimes sabotage also results in creating faults. Accumulation of dirt, dust, salt, smoke, etc. on insulator surface in polluted areas may also result in a fault. Wind and ice loading may cause insulator strings to fail mechanically. Deterioration of insulation due to ageing and overloading of underground cables result in short circuits. Some small animals like rats, lizards, enter switchgear to create faults. Faults may occur at different points in a power system.

1.3 TYPES OF FAULTS

Faults are broadly classified as follows:

1. Symmetrical faults
2. Unsymmetrical faults

Symmetrical Faults

In such type of faults all the three phases are short circuited to each other. There may be two situations. In one case, all the three phases are short-circuited to the ground. This type of fault is called ***three-phase to ground (LLLG) fault.*** In the other case, all the three phases are short-circuited without involving the ground. This type of fault is called ***three phase short-circuit (LLL) fault.*** Faults of the type LLL and LLLG are balanced and symmetrical in the sense that the system remains symmetrical, that is, balanced even after the fault. A three-phase short circuit occurs rarely, but it is the most severe type of fault involving largest currents. For this reason, the balanced short-circuit calculations are performed to determine these large currents to be used to determine the rating of circuit breakers.

Unsymmetrical Faults

The term unsymmetrical fault is used to mean unbalanced conditions. It is a connection or situation which causes an unbalance among the three phases.

An unsymmetrical *shunt fault* is an imbalance between phases or between phase and ground. A *series fault* is an imbalance in the line impedances. It does not involve any connection between lines, or between line and ground at fault point.

Shunt faults in a three-phase system can be classified as follows:

1. Single line-to-ground (LG) fault
2. Line to line (LL) fault
3. Double line-to-ground (LLG) fault
4. Three-phase short circuit (LLL) fault
5. Three-phase to ground (LLLG) fault

A ***single-line-to-ground (LG),*** fault occurs when one phase conductor breaks and falls to ground or when one phase conductor comes in contact with neutral conductor.

A ***line to line (LL)*** fault occurs when two phase conductors are short circuited.

A ***double line-to-ground (LLG)*** fault occurs when two conductors fall and connected through ground, or when two conductors come in contact with the neutral of a three-phase grounded system.

Open Conductor Faults

A break in one or two phase conductors in a 3-phase circuit leads to an open-conductor fault. Such faults may also be caused when circuit breakers or isolators open but fail to close one or more phases. Unbalanced currents flow in the system due to opening of one or two phases. This results in overheating of rotating machines.

All these types of faults may occur in the alternators, motors or transformers, or any part of the power system.

Two or more faults may occur simultaneously on a system. Such faults are known as *multiple* or *simultaneous faults*. In such faults, the same or different types of faults may take place at the same or different points of the system.

1.4 EFFECTS OF FAULTS

Faults can damage or disrupt power systems in several ways.

1. At the fault point itself there may be arcing, accompanied by high temperatures and possible fire and explosion. The explosion may take place in equipment, containing insulating oil.
2. Large voltages may be generated which may stress the insulation beyond their breakdown value.
3. Large currents may be produced which may overheat the power system components. Sustained overheating may reduce useful life of the equipment.
4. Large destructive mechanical forces may be developed due to very high currents.
5. Sometimes faults lower the system voltages below their permissible limits.
6. Faults can cause the three-phase system to become unbalanced with the result that the three-phase equipment operates improperly.
7. Sometimes faults block the flow of power.
8. Faults can cause system to become unstable and lose synchronism.

Hence, it is necessary that, upon the occurrence of the fault, the faulty sections should be disconnected as rapidly as possible in order that the normal operation of the rest of the system is not affected. If this is not done, the equipment may be damaged and the power supply is disrupted. The relays should immediately detect the existence of the fault and initiate circuit breaker operation to disconnect the faulty section.

Faults are removed from a system by opening or "tripping" circuit breakers. These are the same circuit breakers used in normal system operation for connecting or disconnecting generators, lines and loads. For emergency operation the circuit breakers are tripped automatically when a fault condition is detected. Ideally, the operation is highly selective; only those circuit breakers closest to the fault operate to remove or "clear" the fault. The rest of the system remains intact.

1.5 COMPONENTS OF A PROTECTION SYSTEM

Components used for power system protection consist of instrument transformers, relays and circuit breakers. The basic function of each device is shown schematically in Fig. 1.1.

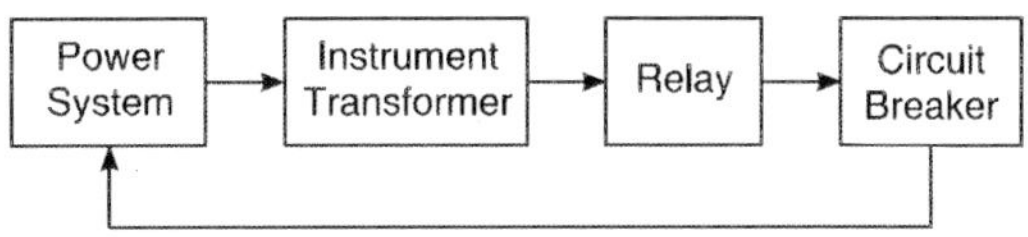

Fig. 1.1 Power system protection components.

The purpose of the instrument transformer is to sense a measurement parameter (normally voltage or current) and send this information to a relay. The relay then processes this information to determine if an abnormal situation exists and acts to achieve coordination with other protection points in the system. If it is determined that action is required at its own point of protection, the relay sends a trip signal to the circuit breaker which then acts to open the power system circuit at that point.

1.5.1 Objectives OF Power System Protection

The important objectives of power system protection schemes are as follows:

1. To detect the fault and to disconnect the faulty section as fast as possible.
2. The normal operation of the rest of the system should not be affected.
3. Since many faults are self clearing, the system should be restored to its original configuration as rapidly as possible.

4. The normal and abnormal conditions should be clearly discriminated so that protective devices do not operate unnecessarily.

Since these are the ideal conditions, it is not possible to satisfy all these conditions at the same time.

1.6 BASIC REQUIREMENTS OF PROTECTIVE RELAYING

An efficient protective relaying should have the following features:

- Reliability
- Selectivity
- Speed
- Simplicity
- Economy

Reliability

Reliability is a measure of the degree that the protective system will function properly in terms of both *dependability* (that is, performing correctly when required) and *security* (that is, avoiding unnecessary operation).

Selectivity (or Discrimination)

Selectivity is the quality whereby a protective system distinguishes between the conditions for which it is intended to operate and those conditions for which it must not operate. In other words, the selectivity of a protective system is its ability to recognize a fault and trip a minimum number of circuit breakers to clear the fault. A relay system should provide maximum possible continuity of supply with minimum system disconnection.

Speed

It is the ability of the protective system to disconnect a faulty system element as quickly as possible with minimum fault time and equipment damage. Therefore, a protective relay should operate at the required speed. The speed should neither be too low, which may result in damage to the equipment, nor should it be too fast which may result in undesired operation during transient faults. The speed of operations also has direct effect on the general stability of the power system. The shorter the time for which a fault is allowed to persist on the system, the more load can be transferred between given points on the power system without loss of synchronism. The operation time for bulk power transfer should not exceed 2.5 cycles for 50 Hz system.

Simplicity and Economy

A protective system should be simple from the point of view of equipment and circuitry. It should also be economical.

1.7 PROTECTIVE ZONES

Relays usually have inputs from several current transformers (CTs), and the zone of protection is bounded by these CTs. The CTs provide the ability to detect a fault inside the zone of protection and the circuit breakers (CBs) isolate the fault by disconnecting all of the power equipment inside the zone. Thus, a zone boundary is usually defined by a CT and a circuit breaker.

In order to cover all power equipment by protective systems, the zones of protection must meet the following requirements :

- All power system elements must be covered by at least one zone. More important elements should be included in at least two zones.
- Zones of protection must overlap to prevent any system element from being unprotected. The region of overlap should be finite and small.

An electric power system is divided into protective zones for

(1) generators (generator - transformer),

(2) transformers,

(3) bus bars,

(4) transmission and distribution circuits, and

(5) loads(motors).

The division is such that zones are given adequate protection while keeping service interruption to a minimum. Each zone has its own protective relays for detecting the existence of a fault in that zone, and each zone has its own circuit breakers to disconnect that zone from the rest of the system. Therefore, a protected zone can be defined as the portion of the power system protected by a given protective system or as part of the protective system. A single-line diagram of a part of a power system with its zones of protection is shown in Fig. 1.2.

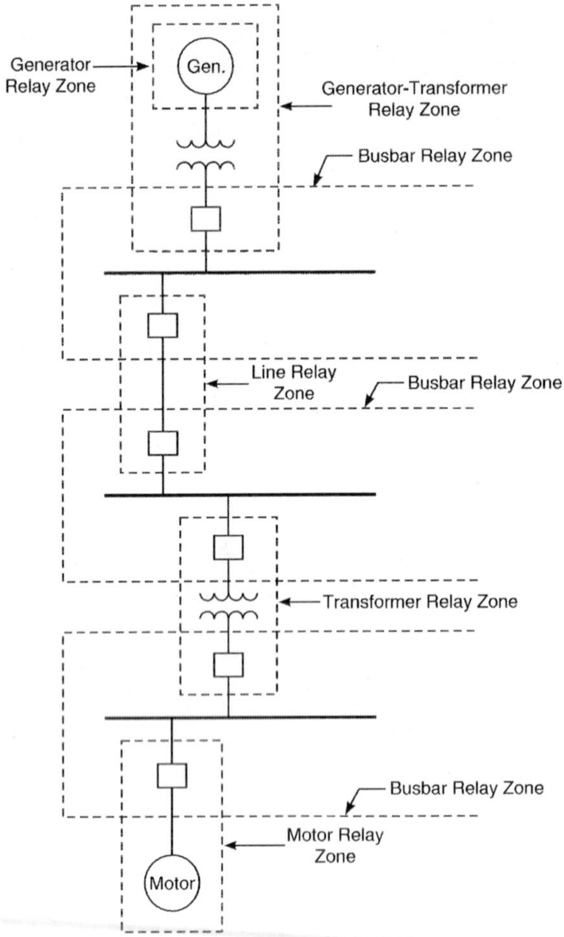

Fig. 1.2 Typical zones of protection in a part of an electric power system

Primary Protection

The protection provided by each zone to its element(s) is known as ***primary protection***.

Backup Protection

In case of failure of primary protection, a backup protection should be provided to remove the faulted part from the system. The primary protection is the first line of defence and the backup protection is the second line of defence. In general, the backup protection is provided for possible failure in primary relaying systems and circuit breakers.

Causes of failure of the primary protection

- Failure of relay itself
- Failure of auxiliary devices
- Loss of dc control supply
- Failure of current and voltage transformers and their circuits

Causes of a circuit breaker failure

- Failure of main contacts to interrupt
- Mechanical failure of tripping mechanism
- Open or short-circuit trip coil
- Loss of dc supply

Therefore, any backup protection must provide both relay backup and circuit-breaker backup.

In general, there are three kinds of backup relays.

1. Relay backup
2. Circuit breaker backup
3. Remote backup

Relay backup

These relays trip the same circuit breaker if the main relay fails.

Circuit breaker backup

These relays open the next nearest circuit breakers on the same bus if one of the local circuit breakers fails to open.

Remote backup

These relays operate from the next station, in the direction towards the source to backup both relays and circuit breakers. Remote backup relaying avoids the simultaneous failure of both the primary and backup relaying. It is to be noted that remote backup is inherently slow. It usually disconnects greater portion of the power system than is necessary to remove the faulty part.

If faster backup relaying is necessary, the backup relaying may be located in the same location as the primary relaying to provide "local backup". Therefore, in local back-up relaying, fault is cleared locally in the same station where it has taken place. However, the local backup relaying should be as completely separated from the primary relaying as is possible.

The measures taken to provide backup protection vary widely depending on the value and importance of the installation and the consequence of failure.

1.8 CLASSIFICATION OF PROTECTIVE RELAYS

Protective relays can be classified in various ways depending on their construction, function, etc.

1. Classification of Protective Relays Based on Technology

Protective relays may be classified into the following categories depending on the technology used for their construction and operation:

(a) Electromagenetic relays

(b) Static relays

(c) Microprocessor based relays

2. Classification of Protective Relays Based on Their Function

Protective relays can be classified into the following categories, depending on the duty they are required to perform:

(a) Overcurrent relays

(b) Undervoltage relays

(c) Impedance relays

(d) Under frequency relays

(e) Directional relays

1.9 CLASSIFICATION OF PROTECTIVE SCHEMES

The most commonly used protective schemes are as follows:

(a) Overcurrent protection

(b) Distance protection

(c) Pilot relaying

(d) Differential protection.

1.9.1 Overcurrent Protection

A protective relay which operates when the load current exceeds a preset value (called the pick up current) is known as overcurrent relay. A scheme in which overcurrent relays are used for protection is called an overcurrent scheme.

1.9.2 Distance Protection

A distance relay responds to input quantities as a function of the electrical distance between the relay location and the fault place. Distance protection is used for the protection of high voltage transmission or subtransmission lines.

1.9.3 Pilot Relaying

The term ***pilot relaying*** essentially means remote control of circuit breakers. In this type of relaying the decision to open or close a circuit breaker is made geographically remote from the location of the circuit breaker. Pilot relaying is the best type of line protection. It is a high-speed protection.

1.9.4 Differential Protection

A relay that operates when the phasor difference of two or more similar electrical quantities exceeds a predetermined value. Differential protection is used for protection of generators, transformers, busbars, large motors etc.

1.10 INSTRUMENT TRANSFORMERS

Usually protective relays are not directly connected to the system. To provide low level equivalents of system currents and voltages relays are connected through current and/or voltage transformers. These are known as instrument transformers. The main functions and advantages of instrument transformers are as follows:

1. Instrument transformers are used to reduce high voltages and currents to useable values for measuring and protection purposes.
2. Measuring and protection equipments need not be insulated for high voltages.
3. Working at reduced values of currents and voltages is safer.
4. Simple and reliable protection and measuring circuitry.
5. Reduction in currents and/or voltages leads to lesser insulation requirement and current capacities of conductors.
6. With the use of instrument transformers the protection and measuring systems become compact and economical.

Following important considerations are to be taken into account for proper application of instrument transformers.

- Ratio in terms of primary and secondary voltages or currents (that is input to output ratio)

- Accuracy
- Type of requirement (measuring or protective)
- Type of insulation
- Thermal ratings (continuous and short time)
- Impulse current/voltage level
- Service conditions and connections
- Mechanical construction.

1.11 CURRENT TRANSFORMER (CT)

Current transformers are used to step down current from a high value to a low value. The primary winding of the current transformer is connected in series with the load and carries the system current. The secondary winding is connected to the measuring or protective circuit (relay).

The CTs usually used for protection purposes are different from CTs used only for measuring purposes. CTs used in measuring purposes require accuracy over the normal working range of the system current. CTs used for protection purposes need to be accurate for upto several times the normal system current (primary current) Current transformers used for protection usually have a secondary circuit rating of 5A or 1A.

Any abnormal increase in primary current due to some fault saturates the core. Increase in the burden (above VA capacity) of the CT can also cause such a condition.

Deviation of actual current ratio from nominal current ratio is observed as the CT core enters saturation. This is known as ratio error and is defined as follows:

$$\text{Ratio Error} \triangleq \frac{\text{nominal ratio-actual ratio}}{\text{actual ratio}}$$

$$= \frac{N - \left(\dfrac{I_p}{I_s}\right)}{\left(\dfrac{I_p}{I_s}\right)}$$

$$= \frac{NI_s - I_p}{I_p}$$

where, N = nominal ratio = $\frac{\text{rated primary current}}{\text{rated secondary current}}$

I_p = actual primary current

I_s = actual secondary current

Deviation in phase angle of the reversed secondary current phasor with primary current phasor is also observed. This is called *phase angle error.*

Usually the ratio error is largely dependent on wattful component of the exciting current and the phase angle error depends on the magnetizing component of the exciting current. Phase angle errors are not of much concern in case of CTs. Usually lower power factor loads (relay coils) are connected which compensate the phase angle errors. Phase angle errors can be tolerated in protective relaying.

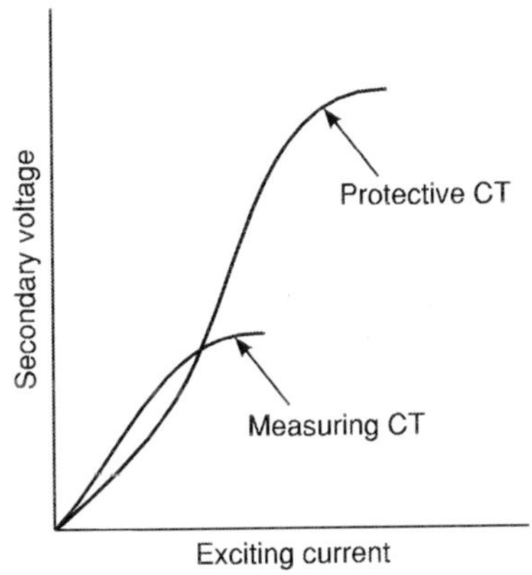

Fig. 1.3 Magnetization curves for protective and measuring CTs

In order to reduce these errors low reluctance core is provided and designed for low values of core losses. The flux density is to be kept as low as possible. Magnetization characteristics of measuring and protective CTs are shown in Fig 1.3.

Nickel-iron alloys, cold rolled grain-oriented silicon steel cores, are widely used because of their high permeability, high saturation level, and draw small exciting current. Sometimes for practical purposes ratio error can be compensated by reducing the turn ratio by one or two as required. Generally ratio error is corrected using ratio-correction factor curves. Ratio-correction-factor curves are plotted against multiples of rated primary or secondary current for different constant values of burdens.

Current transformers can be classified into two categories on the basis of their construction.

1. Bar primary ring core type CT
2. Wound primary CT

For protection purposes bar primary CTs are widely used due to their large withstand capacity for large electromagnetic forces developed due to short circuits on the system.

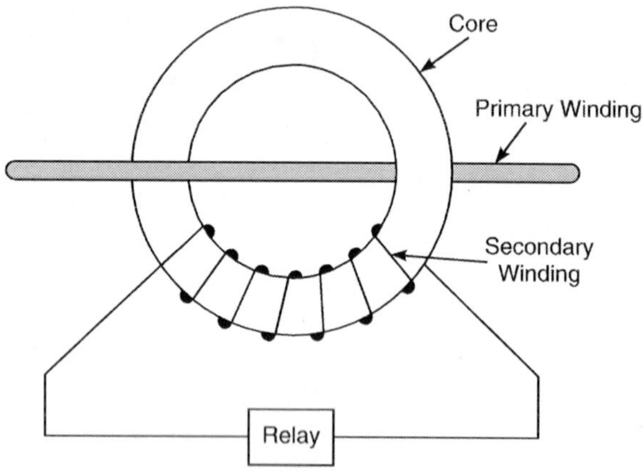

Fig. 1.4 Bar primary ring core type CT

A bushing CT is a special type of bar primary CT placed over an insulator bushing enclosing a straight conductor. It is simple in construction, provide linear input/output characteristics and economical. Sometimes this type of transformer is built into various equipments.

Core having air gaps provide saturation free transformation. These are called linear cores. Non ferrous cores are also used because of their linear characteristics. These are air or plastic cored CTs and are known as linear couplers. These types of core constructions provide error free operation even for large transient dc components of fault current.

The CT must never be operated with secondary winding open-circuited. If the secondary winding is open circuited, the whole of the primary current becomes the magnetizing current. Since the secondary current is zero, the secondary ampere-turns are also zero and there is nothing to balance the primary ampere-turns. This gives rise to a very high flux in the core causing magnetic saturation. Consequently a large voltage is induced in the secondary winding which may damage the insulation. For momentary open circuits the CT core must first be demagnetized before being used.

CT Burden

The external load applied to the secondary winding of the CT is called burden. The burden is expressed in terms of impedance of the load and is expressed as volt-amperes.

1.12 VOLTAGE TRANSFORMER (VT) OR POTENTIAL TRANSFORMER (PT)

Voltage transformers are used to step down the system voltage to a level enough to match the voltage relays of protective or measuring equipment. The voltage transformer is similar in construction and connections to the power transformers, the difference being the voltage transformer is rated in terms of maximum VA output (burden) it can deliver without exceeding the specified limits of the error and the power transformer rating specify its secondary output (kVA) it can deliver without exceeding the specified temperature rise limit. Also as the output is low it is smaller in size and needs less cooling arrangements.

Ratio and phase angle errors occur in a voltage transformer as in case of current transformers but in general are of lesser importance. Both the errors depend on the following

- Primary and secondary winding resistances due to their voltage drops.
- Excitation current of the potential transformer
- Burden on the secondary winding
- Power factor of the burden (including secondary winding)

To reduce the errors in voltage transformers

- The resistance and the leakage reactance of the windings should be as low as possible.
- The reluctance of the transformer core should be as low as possible.
- The flux density in the core should be low.
- Turn compensation should be done if necessary.

Two types of constructions of voltage transformers are used for protective purposes.

(i) Electromagenetic type
(ii) Capacitance type

In electromagnetic type VT the primary is connected to the system directly between the two phases or between a phase and ground and the secondary is connected to the protective circuit. In VT capacitor voltage divider type VT the capacitance divider is connected between a phase and ground of the circuit and the output is taken between any desired point on voltage divider and ground and fed to the protective circuit.

Electromagnetic type construction is preferred for voltages up to 66 kV. Above 66 kV, generally capacitance voltage divider type VTs are used because of economy. A capacitor voltage type divider VT with error compensation is shown in Fig 1.5.

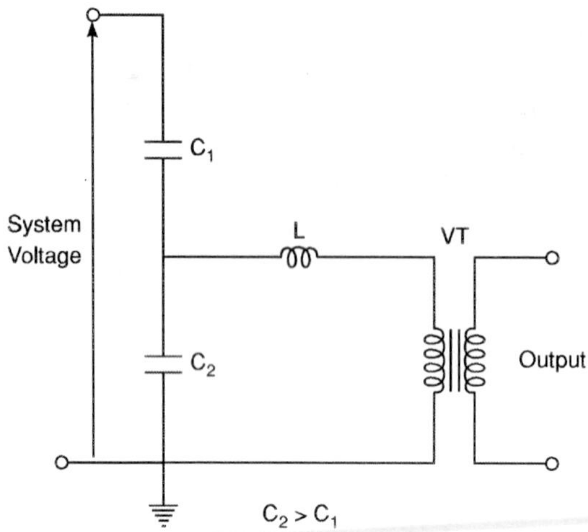

Fig. 1.5 Capacitor voltage divider VT with compensation.

In order to reduce ratio and phase angle errors due to variation in burden etc. the inductor L is connected and tuned to the VT. The inductor is tuned such that it resonates with capacitors at system frequency. When two or more lines are running from a common bus a single VT may be connected to the bus to supply the protective relaying equipment of all the lines .For this purpose and voltages above 66 kV electromagnetic type VTs are cascaded.

Capacitor type VTs are less reliable for high speed relaying schemes. They also have a tendency to introduce harmonics in the output voltage and have an inferior transient performance as compared to electromagnetic type VTs.

EXERCISES

1. Explain the nature, causes and consequences of faults.
2. What are the main objectives of power system protection ?
3. What are different types of faults ?
4. What is a zone of protection ? Discuss various zones of protection for a power system with the help of a line diagram.
5. What are primary and backup protections ? What are various kinds of backup protections ?
6. What are the basic requirements of protective relaying schemes ?
7. Explain the terms
   (a) Selectivity
   (b) Reliability
   of a protective relay.
8. Classify protective relays on the basis of the technology and function.
9. What are instrument transformers ? Give their advantages and uses.
10. What are ratio and phase angle errors in CTs ? Give causes and remedies for them
11. What are the advantages and disadvantages of electromagnetic type VTs over capacitor voltage divider type VTs ?

❑❑❑

2

ELECTROMAGNETIC AND ELECTRO-THERMAL RELAYS

2.1 PRINCIPAL TYPES OF ELECTROMAGNETIC RELAYS

The electromagnetic type relays depend upon the production of electromagnetic forces on movable members, which provide switching action by physically opening or closing sets of contacts.

There are two main types of electromagnetic relays:

(1) Attracted armature type

(2) Induction type

Attracted armature type relays are compact, robust and reliable. They are also faster than induction disc and cup type relays. Their VA burden is also low but are affected by transients.

2.2 ATTRACTED ARMATURE TYPE RELAYS

These type of relays include the following

(a) Plunger type relay

(b) Balanced beam type relay

(c) Polarized relay

(d) Hinged armature type relay

2.2.1 Plunger- Type Relay

A typical electromagnetic plunger- type relay is shown in Fig. 2.1. It consist of cylindrical coils on a magnetic structure and a centre plunger.

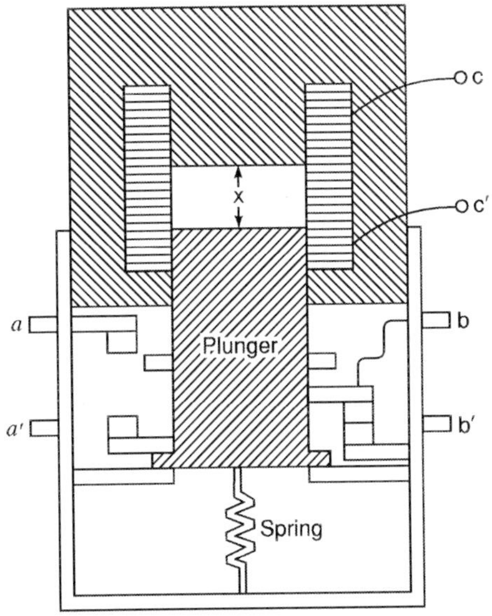

Fig 2.1 An electromagnetic plunger-type relay

The plunger is movable. When the coil is energized, magnetic force will attract the plunger upwards closing the air gap x. The spring pulls the plunger down when the coil is deenergized. Under normal conditions the coil *cc′* is not energized, the contact *aa′* are open and contacts *bb′* are closed.

The plunger is set to move upward (to close the contacts *aa′*) when the current applied to the coil exceeds a certain value. The moving force is proportional to the square of the current in the coil. The threshold value of current at which the plunger starts moving upward is called pickup current of the relay unit.

Plunger type relays can operate on dc as well as ac and can be designed to respond to voltage as well as current if the coil impedance is high. Typical operating times are 5 to 50 ms with longer time near the threshold value of pick up current.

2.2.2 Balanced-Beam Relay

The balanced beam relay operates on the balance principle by comparing two quantities, for example, two currents or one current and one voltage. Fig. 2.2 shows an electromagnetic attraction *balanced-beam type* relay. It is the general form of the relay comparing two currents. The beam (armature) is attracted by the electromagnets which are operated by the appropriate parameters, usually two currents or one current and one voltage.

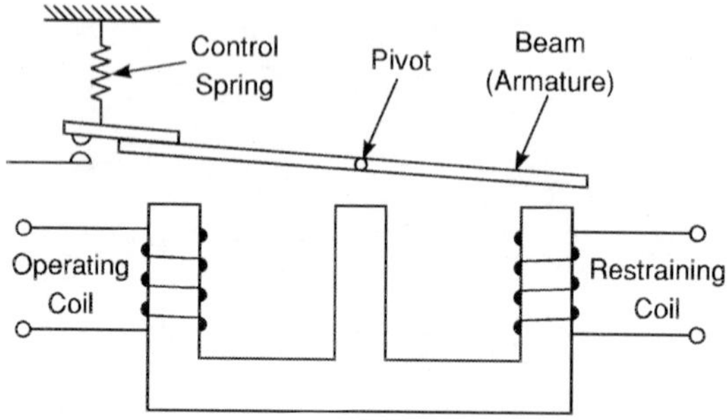

Fig. 2.2 Balanced beam relay

The beam is given a slight mechanical bias to keep the contacts open, except when operation is required. This is achieved by a control spring or weight adjustment. Suppose that the two coils have equal number of turns. The relay will operate when

$$I_a^2 = I_b^2 + K$$

where, I_a = current in the operating coil

I_b = current in the restraining coil

K = a constant

2.2.3 Polarized Relays

A relay whose operation depends upon the direction of current or polarity of the voltage is called a polarized relay.

Polar relays use direct current applied to a coil wound around the hinged armature on a magnetic structure. A permanent magnet used for polarization, produces flux in addition to the main flux making it sensitive to the polarity of the input. That is, it operates only when the input is of correct polarity.

Two type of constructions used are:

1. Moving Coil Type
2. Moving Iron Type

Figure 2.3 shows a moving coil type polarized relay. The coils are wound on a cylindrical former and suspended to have only axial movement. An operating torque is produced due to the interaction of field produced by the current in the coil and the permanent magnet. This torque is proportional to the current flowing in the coil.

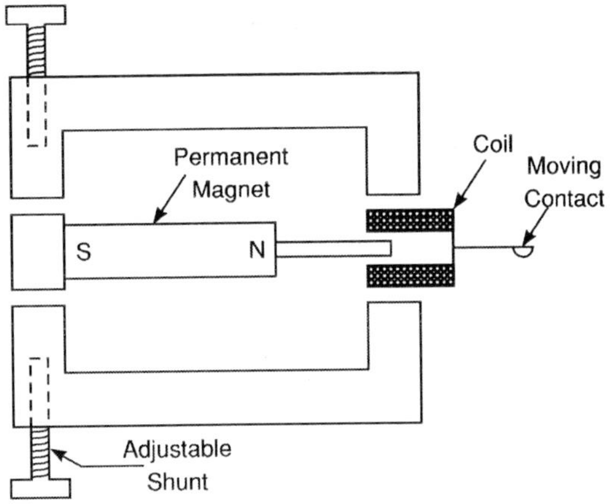

Fig. 2.3 Polarized moving coil relay

Sometimes two nonmagnetic spacers, located at the rear of the magnetic frame are bridged by two adjustable magnetic shunts. This arrangement enables the magnetic flux paths to be adjusted for pickup and contact action.

This type of relay can also be used with ac with suitable rectifiers. Moving coil relays are most sensitive among the electromagnetic relays and have a small VA burden.

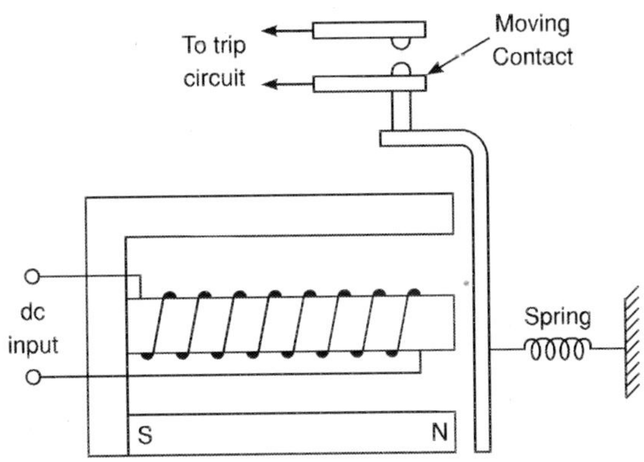

Fig. 2.4 Polarized moving iron relay

Construction similar to indicating type instrument is also used for moving coil relays. This type of construction is slower and less sensitive as compared to axial type construction.

Figure 2.4 shows a polarized moving iron relay. The current carrying coil is wound on a stationary core and a permanent magnet used for polarization is put as shown in Fig. 2.4. The contacts are closed as the currents of correct polarity exceeds the pickup value.

2.3 ELECTROMAGNETIC INDUCTION TYPE RELAYS

In these types of relays, the operating torque is produced by induction principle. Therefore, these relays can only be used on ac systems. These are basically a two-quantity relays which can be easily adopted for single quantity operation. Torque is produced in these relays when one alternating flux reacts with the current induced in the rotor by another alternating flux displaced in time and space but having the same frequency. Depending upon the type of rotor being used these relays are classified as

(a) induction disc type

(b) induction cup type

The induction disc type relays are categorized as ***shaded-pole type*** and ***wattmetric type*** induction relays.

2.3.1 Shaded Pole-Type Induction Disc Relay

Figure 2.5 shows the shaded pole-type induction disc relay. This type of construction is used for a single quantity relay. The actuating quantities may be current or voltage.

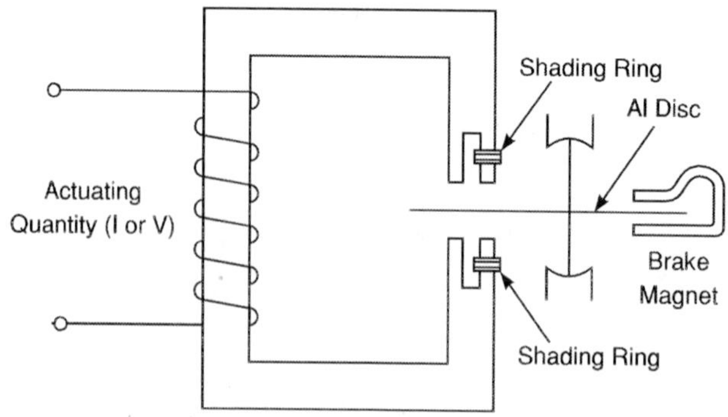

Fig. 2.5 Shaded-pole type induction disc relay

The shaded pole relay consists of an operating coil carrying the current proportional to the system current. The air gap flux produced by this current. The main flux is split into two fluxes displaced in time and space with the help of a shading ring. The air gap flux of the shaded poles, lags behind the flux of the unshaded poles. The shading ring is made of copper that encircles part of the pole face of each pole at the air gap. The moving part is a thin circular metallic disc normally of aluminium or copper (non-magnetic conductor). This disc is pivoted smoothly at its centre and is free to rotate.

2.3.2 Wattmetric Type Induction Disc Relay

A thin aluminium disc is free to rotate between two fixed terminated electromagnets. These electromagnets are energized by two input quantities A and B. If both the inputs are derived from the same source, sufficient phase displacement is produced between the inputs. In such type of cases the torque is proportional to the square of the actuating current. The aluminium disc is placed between these two electromagnets so that it cuts the fluxes produced by both the electromagnets. To restrain the movement of the disc a spring is attached.

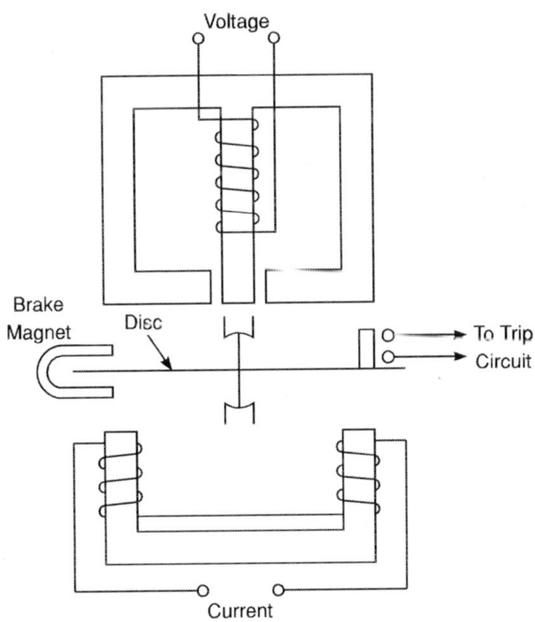

Fig. 2.6 Wattmetric type induction disc relay

Eddy current braking is used to prevent overrun of the disc due to disc inertia. A permanent magnet is used for this purpose. The braking torque is proportional to the speed of the disc.

The time of operation of the relay depends upon the distance by which the moving contact of the relay travels before closes the contacts. Operating time is adjusted by changing the position of backstop on which disc rests under normal conditions. Induction type construction provides a slow operation and can be used where time delay is required.

2.3.3 Induction Cup Type Relay

Figure 2.7 shows an induction cup type relay. It consists of a thin walled aluminium cylinder with one end closed with a cup. There is a fixed inner steel core at the centre of a square electromagnet. The thin aluminium cylinder is free to rotate in any annular air gap between the pole faces of electromagnets and the central steel core. The steel core provides low reluctance path to the magnetic fluxes. Since the operation of this relay is similar to that of an induction motor with salient poles for the stator windings, this relay is called induction cup type relay.

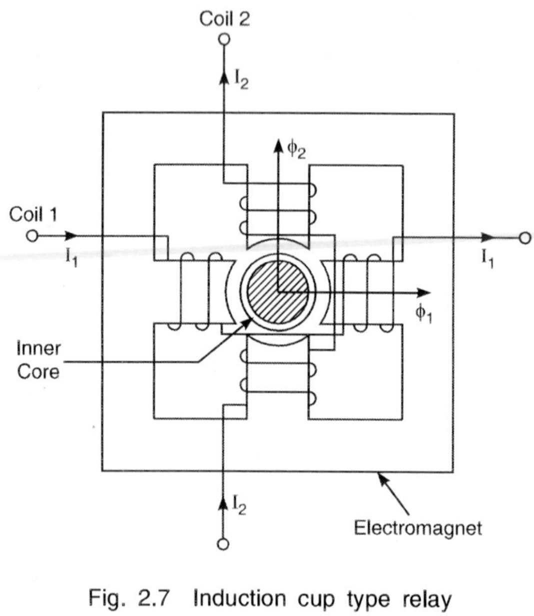

Fig. 2.7 Induction cup type relay

The stator poles carry two pairs of coils. They are excited by the input quantities to produce a rotating field. The rotating field induces current in the cup causing it to rotate in the direction of the rotating field. The spindle of the cup carries an arm which closes contacts. A spiral spring is used to provide a resetting torque. The rotation of the cup is limited to a few degrees to connect the moving and fixed contacts. The operating torque is the product of the two operating quantities and the sine of the angle between them.

$$\tau = kAB \sin \theta$$

where A and B are the operating quantities

The induction cup type relay has several advantages over the induction disc type relay.

1. Since the rotating parts of the induction cup relay are of low inertia, it is capable of high-speed rotation. Therefore, it can be used for function requiring instantaneous operation. The operation time is less than 0.01 second.
2. It has high sensitivity and minimum magnetic leakage.
3. It has the ability to operate on low voltages.
4. There is no current or voltage creep. That is, if there is no voltage or current, the rotor does not move.
5. It can be made to have linear operating characteristics and a very high ratio of resetting to operating values.
6. This type of relay is best suited where normal and abnormal conditions are very close together.
7. These relays impose less burden on the main circuits.
8. Transient effects can be minimized with proper designs.

The multiplicity of poles permits the measurement of more than one electrical quantity. Different combinations of input quantities can be used for different applications, system voltages or currents, or network voltages.

Induction cup relays are most suitable for distance and directional relays.

2.4 THEORY OF INDUCTION RELAY TORQUE

Let ϕ_1 and ϕ_2 be the sinusoidal fluxes produced in the induction relays. If θ is the phase angle between them

$$\phi_1 = \phi_{1m} \sin \omega t$$

$$\phi_2 = \phi_{2m} \sin (\omega t + \theta)$$

The flux ϕ_2 leads ϕ_1 by θ.

In case of shaded-pole construction ϕ_1 is produced by the shaded pole and ϕ_2 by the unshaded pole. In wattmetric type construction ϕ_1 is produced by the upper electromagnet and ϕ_2 by the lower electromagnet. In cup-type construction ϕ_1 and ϕ_2 are produced by pair of coils. The theory of torque production given below is valid for both the disc type and cup type induction relays.

The two fluxes ϕ_1 and ϕ_2 will induce voltages in the rotor. The voltages induced in the rotor are given by

$$e_1 \propto \frac{d\phi_1}{dt}$$

$$\propto \frac{d}{dt}[\phi_{1m} \sin \omega t]$$

$$\propto \phi_{1m}\, \omega \cos \omega t$$

$$e_2 \propto \frac{d\phi_2}{dt}$$

$$e_2 \propto \frac{d}{dt}[\phi_{2m} \sin (\omega t + \theta)]$$

$$\propto \phi_{2m}\, \omega \cos (\omega t + \theta)$$

These voltages will circulate eddy currents in the rotor. If the rotor is assumed to be noninductive, these currents will be in phase with their respective voltages. Assuming same resistance to the flow of eddy currents

$$i_1 \propto e_1$$

$$i_2 \propto e_2$$

$$i_1 \propto \phi_{1m} \cos \omega t$$

$$i_2 \propto \phi_{2m} \cos (\omega t + \theta)$$

The flux ϕ_1 will interact with eddy current i_2 and ϕ_2 will interact with i_1. Since ϕ_2 is leading ϕ_1, the torque due to ϕ_2 and i_1 will be taken as positive and the torque due to ϕ_1 and i_2 will be negative.

The resultant torque is

$$\tau \propto \phi_2 i_1 - \phi_1 i_2$$

$$\phi_2 i_1 - \phi_1 i_2 = [\phi_{2m} \sin (\omega t + \theta)]\, \phi_{1m} \cos \omega t - [\phi_{1m} \sin \omega t]\, [\phi_{2m} \cos (\omega t + \theta)]$$

$$= \phi_{1m}\, \phi_{2m} \sin \theta$$

$$\therefore \quad \tau \propto \phi_{1m}\, \phi_{2m} \sin \theta$$

2.5 ELECTRO-THERMAL RELAYS

The operation of thermal relays depends upon the electro-thermal effect of the actuating current. The actuating current produces the temperature rise. This is detected by a thermal relay.

A simple thermal relay consists of a bimetallic element. The bimetallic element consists of two elements of different thermal linear expansions (Fig. 2.8 a) joined together. During heating the expansion of one strip is more than the other. Consequently, strips are bent to close the contacts (Fig. 2.8 b). The spiral construction is used to save space and to make it more sensitive.

Sometimes unimetallic strips in V-shapes (Fig. 2.8 c) are also used. On being heated, the strip expands to close the contacts.

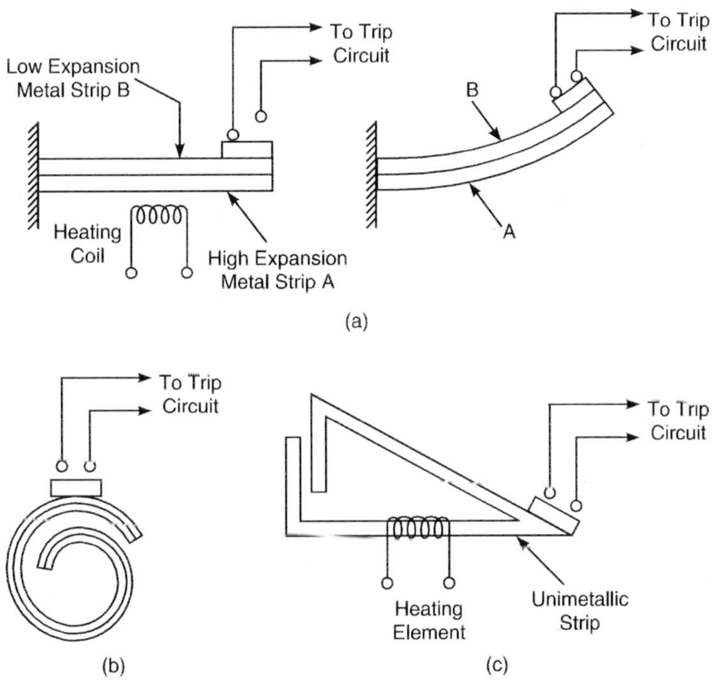

Fig. 2.8 Thermal relays
(a) Bimetallic linear type, (b) Bimetallic spiral type, (c) Unimetallic type

Simple thermal relays are used in protection against overloads of small motors.

In case of a fault, negative sequence currents flow in three phase systems. Thermal relays can also respond to unbalanced three-phase currents, which cause a temperature rise due to their negative sequence component.

In the protection of large alternators thermal relays are used in the form of thermocouples and resistance temperature detectors. These elements are placed in the stator slots. Such an element forms an arm of a balancing bridge. Under normal operating conditions the bridge is balanced. When the current exceeds a prescribed limit the bridge becomes unbalanced and the relay is energized to trip the circuit breaker.

EXERCISES

1. What are the different types of electromagnetic relays ? What are their fields of application ?
2. Explain with the help of a diagram the working principle of an electromagnetic plunger type relay.
3. Explain the working of a balanced beam relay.
4. What is a polarized relay ? Explain with neat sketches the working principles of moving coil type and moving iron type polarized relay.
5. Draw a neat sketch of an induction disc relay and explain its principle of operation.
6. Describe the construction and working of an induction cup type relay.
7. What are the advantages of induction cup type relay over induction disc type relay ?
8. Derive an expression for torque produced by an induction relay.
9. Discuss the working principle, types and applications of thermal relays.

❑❑❑

3

STATIC RELAYS

3.1 INTRODUCTION

A static relay is an electrical relay in which the function of measurement or comparison of electrical quantities is performed by a static network designed to give an output signal when a threshold condition is achieved. The static network may comprise of electronic, magnetic, optical or other components which are static in nature i.e., without any mechanical motion. The output signal operates a tripping device or some alarm system which may be either be electronic or electromagnetic.

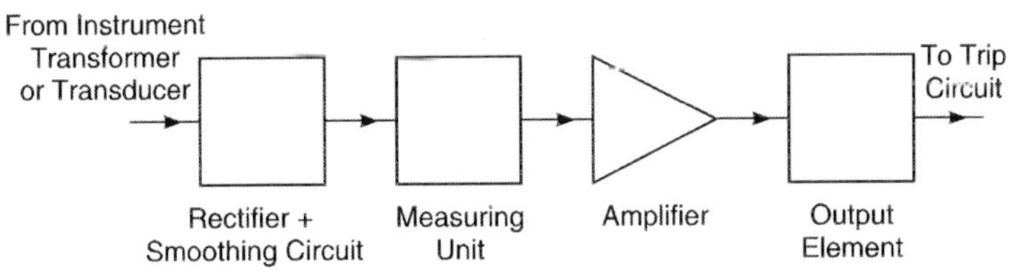

Fig 3.1 Block diagram of a static relay

Figure 3.1 shows a simplified block diagram of a static relay. The system variables such as voltage and current etc. or their derived values are obtained from CTs, PTs or transducers. The output of CTs , PTs or transducers is fed to the rectifier. The rectified smoothed output is now fed to the measuring unit which comprises of comparators, level detectors, logic circuits and filters.

An output signal is actuated when the input to the measuring unit reaches the threshold value. This output signal is amplified and given to the output element. The output element may also provide some intentional time delay if necessary. Comparator may either be an amplitude comparator or a phase comparator, Level detectors can be Schmitt trigger circuit or an OPAMP based circuit. This amplified signal sounds an alarm or energizes the trip coil of the circuit breaker.

3.1.1 Advantages and Limitations of Static Relays

Static relays possess the following *advantages*:

(i) Low burden on current and potential transformers. They consume less power and generally work on auxiliary dc power source.
(ii) Unconventional relaying characteristics can easily be obtained.
(iii) Absence of any mechanical inertia or vibration.
(iv) Fast response.
(v) Long life, low maintenance due to absence of moving parts.
(vi) Compact size, Modular construction is easy.
(vii) Suitable for frequent operation.
(viii) Quick reset action and free from overshoot.
(ix) Greater sensitivity as desired amplification is provided easily.

The *limitations* of the static relays are:

(a) Static relays are temperature sensitive. Temperature compensation is required.
(b) Prone to damages caused by low power transients. Filters and shields add to cost.
(c) Static relays generally need an auxiliary power source for their operation.
(d) Very limited overloading capacity.

3.2 RELAY COMPARATORS

In the event of a fault on a system, the magnitude of voltage and current, and phase angle between voltage and current may change. These quantities during faulty conditions are different from the quantities under normal operating conditions, In general, the task of a relay is to detect the change between normal and faulted conditions and send a signal when the fault occurs. The relay measures different functions of applied quantities and compares between two or more different inputs, or simply compares one measured quantity with a standard (predetermined) value. Relay comparators can have any number of input signals. However, we shall consider here only the two-input comparator. There are various types of comparators. The two most common types of comparators are amplitude comparators and phase comparators.

In a static relay it is necessary to convert the input to the comparator to quantities of the same type, namely, voltages or currents, which can then be compared.

Amplitude Comparator

An amplitude comparator compares the magnitudes of two input quantities, irrespective of the angle between them. One of the input quantities is an operating quantity and the other is a restraining quantity. When the amplitude of the operating quantity exceeds, the amplitude of the restraining quantity, the relay sends a tripping signal.

Phase Comparator

A phase comparator compares two input quantities in phase angle, irrespective of their magnitudes and operates if the phase angle between them is less than or equal to 90°.

The electromagnetic relays that operate inherently as amplitude comparators include balanced beam, hinged armature, plunger, and induction disc relays with shaded-pole driving magnets. The electromagnetic relays that operate inherently as phase comparators include induction cup, induction dynamometer, and induction magnets. The solid-state relays are capable of producing any desired relay characteristics using amplitude or phase comparison.

3.3 AMPLITUDE COMPARATORS

Amplitude comparators are basically of three types.

(i) Rectifier bridge type
(ii) Instantaneous type
(iii) Sampling type.

3.3.1 Rectifier Bridge Type Amplitude Comparator (Integrating Comparator)

Two sinusoidal quantities S_1 and S_2 to be compared in magnitude are rectified by bridge rectifiers are $|S_1|$ and $|S_2|$. Their difference

$|S_1| - |S_2|$ is obtained by connecting the two rectifier bridges in anti-parallel as shown in Fig. 3.2.

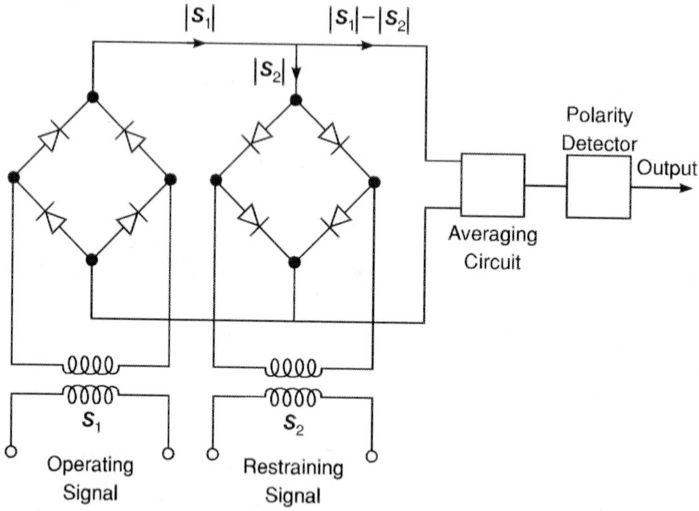

Fig. 3.2 Rectifier bridge comparator with static output circuit

The difference $|S_1| - |S_2|$ is then averaged by an averaging circuit. The average value is then applied to a polarity detector which produces an output if $|S_1| > |S_2|$. Thus, when the operating quantity exceeds the restraining quantity, the relay operates. It is to be noted that S_1 and S_2 could be two voltages or currents. When a voltage and a current is to be compared (as in an impedance relay), one of these should be converted to the form of the other.

The rectifier bridge type comparators are widely used for the realization of overcurrent and distance relay characteristics.

3.3.2 Instantaneous Amplitude Comparators

Instantaneous amplitude comparators are also known as direct amplitude comparators. These can be divided into two sub classes

1. Averaging type instantaneous amplitude comparators

Figure 3.3 (a) shows the block diagram of averaging type instantaneous comparator. The operating signal is rectified and compared with the restraint level derived by rectifying and smoothing the restraining signal. The operating signal sometimes may not be rectified but never smoothed. The output tripping signal is produced when the operating signal exceeds the restraint level. The process is shown in Fig 3.3 (b). Using filter causes delay in the output. Hence, this method is not useful in high speed relay circuits.

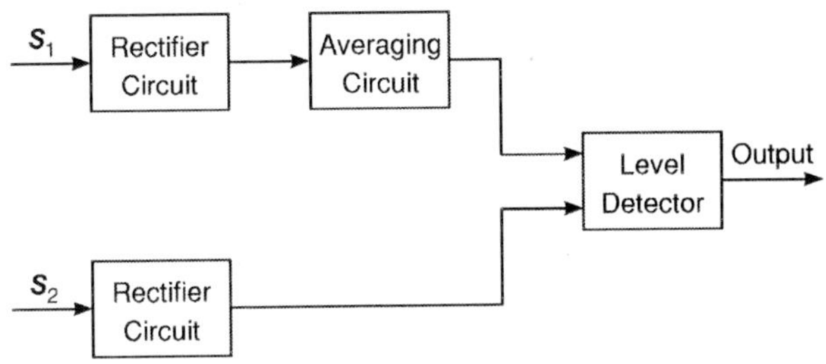

Fig 3.3 (a) Block diagram of averaging type instantaneous amplitude comparator.

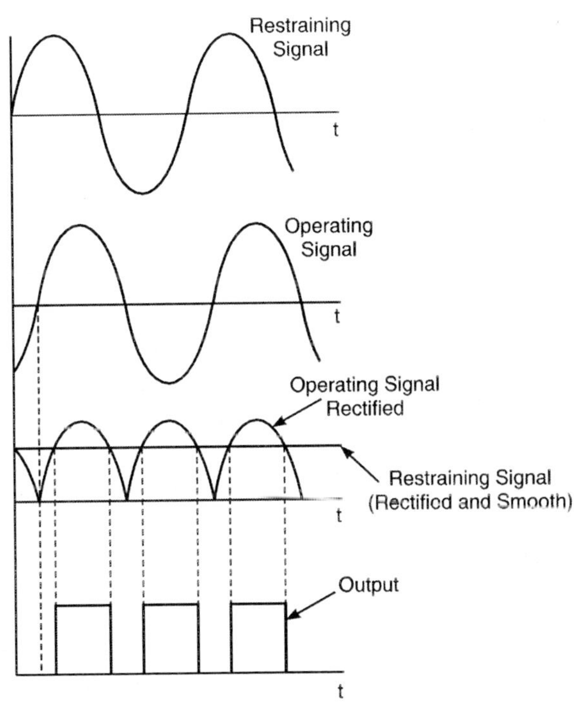

Fig. 3.3 (b) Waveforms of an instantaneous amplitude comparator

2. *Phase splitting type instantaneous amplitude comparator*

Polyphase rectifying methods are used for high speed relay circuits. Various methods of phase splitting are available on the number of phases and amount of smoothing required. Greater the number of phases, greater the smoothing is achieved.

Figure 3.4 shows a single phase input split into three phase output. Phase splitting circuits shift the phases by +120° and −120° in each input, thus giving a 120° shift to each of the components. These ac components are fed to rectifiers giving six phase rectification. The sensor shown may be a polarized relay or a level detector circuit.

As the operating signal exceeds the restrain level the relay operates.

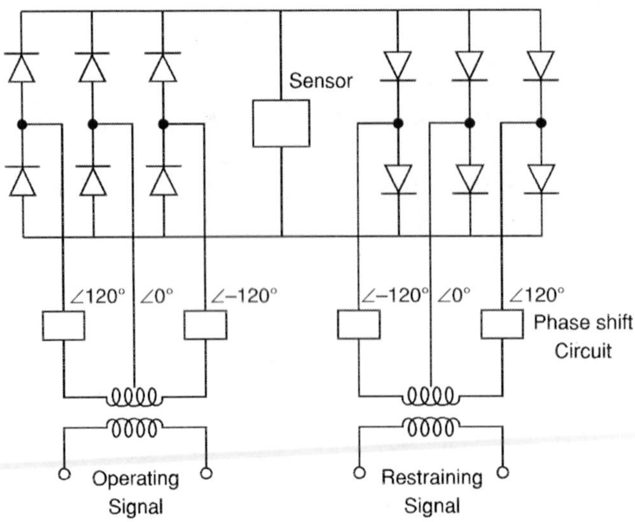

Fig. 3.4 Phase splitting type amplitude comparator

3.3.3 Sampling Comparators

To realize some particular characteristics such as mho and reactance relay characteristics, one input signal at a predetermined point on its waveform is compared with the rectified and smoothed value of the second input. These predetermined points are generally current or voltage zeros, maxima or minima.

3.4 PHASE COMPARATORS

In a static phase comparator, a dc output voltage is obtained, the polarity of which depends upon the phase angle between two sinusoidally varying electrical inputs. This voltage is positive (or in the tripping direction) when the phase angle between the two inputs is in the range $-90° < \phi < 90°$ and negative (or in the blocking direction) when the phase angle is outside the above range. The polarity of the voltage is completely independent of the magnitude of the two inputs.

Basically there are two types of static phase comparators

1. Vector-product type
2. Coincidence type

Vector-Product Phase Comparators

In these type of comparators a dc voltage proportional to the vector product [$\boldsymbol{A} \times \boldsymbol{B} = AB \sin \theta$ or $\boldsymbol{A} \cdot \boldsymbol{B} = AB \cos \theta$] of the alternating quantities is obtained. This type includes Hall-effect phase comparators, and magneto resistivity phase comparators. These relays have been used mainly in the USSR only. At present they are not used and, therefore, will not be discussed here.

Coincidence-Type Phase Comparators

In this type of comparators, a dc output voltage is obtained during the period when the two quantities or their resolved components are all of the same polarity. These can be of either the direct acting (instantaneous) or integrating types.

In a coincidence-circuit type phase comparator, the period of coincidence of similar polarities of two input signals is measured and compared with the predetermined angle, usually 90°. Figure 3.5 shows two input signal having a phase difference of ϕ. The period of coincidence is represented by an angle ψ. It is seen that $\psi = 180° - \phi$ If ϕ is less than 90°, ψ will be greater than 90°.

The relay trips when $\phi < 90°$. This, the phase comparator is designed to send a trip signal when $\psi > 90°$. The period of coincidence can be measured in a number of different ways. Some of the important techniques are as follows:

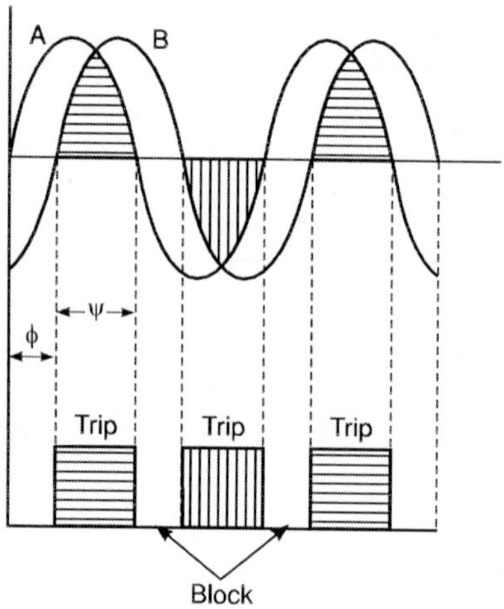

Fig 3.5 Coincidence principle for phase comparators

(a) Phase-splitting type phase comparators

(b) Integrating type phase comparator

(c) Rectifier-bridge type phase comparator

(d) Phase comparator based on pulse signals.

The principles of operation of these comparators are as follows:

3.4.1 Phase Splitting Type Phase Comparator

In this method, the two sinusoidal signals S_1 and S_2 whose phase is to be compared, are split into two components shifted in phase by $\pm 45°$ to produce $S_1 \angle 45°$, $S_1 \angle -45°$ and $S_2 \angle 45°$, $S_2 \angle -45°$ as shown in Figures 3.6 (a) and (b). The four components are fed to an AND circuit as shown in Fig. 3.6 (c). The tripping occurs when all the four signals, become simultaneously positive at any time during the cycle. An AND gate is used as, a coincidence detector. The coincidence of all the four signals occurs only when the phase difference ϕ between S_1 and S_2 is less than 90°. The full range of operation is

$$-90° < \phi < 90°$$

This technique is based on direct comparison.

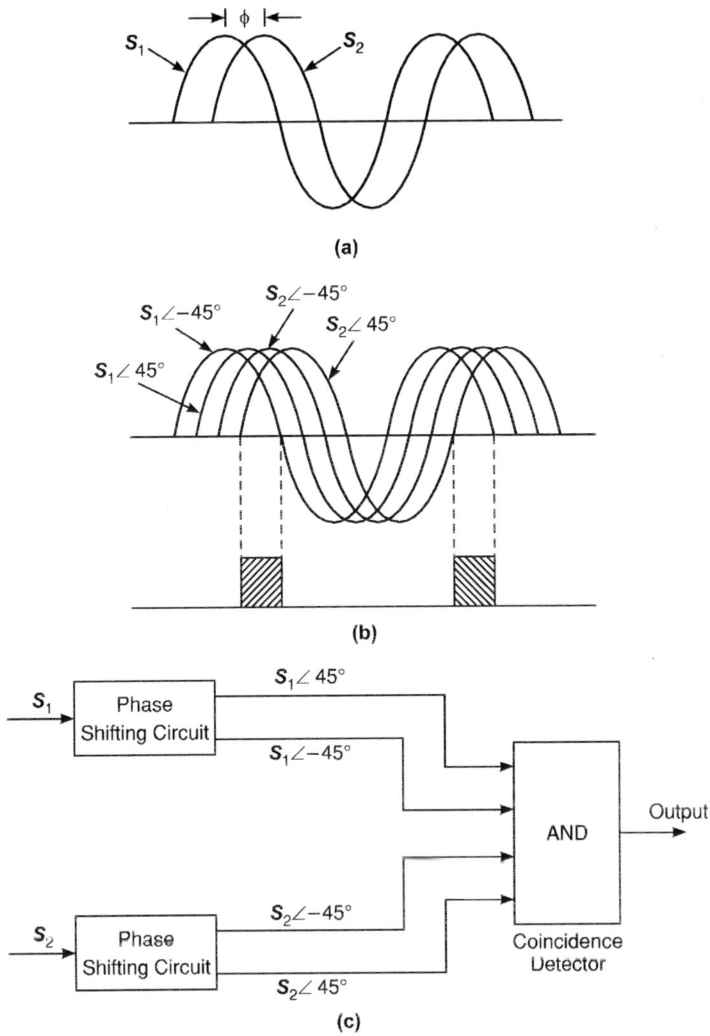

Fig. 3.6. Phase splitting technique for phase comparators.

3.4.2 Rectifier Bridge Type Phase Comparator

Figure 3.7 shows a rectifier bridge type phase comparator. The output voltage appears across the two equal output resistors at terminals a and b. The two inputs S_1 and S_2 are applied through the transformers T_1 and T_2. Transformer T_1 is centre-tapped at the secondary winding. To

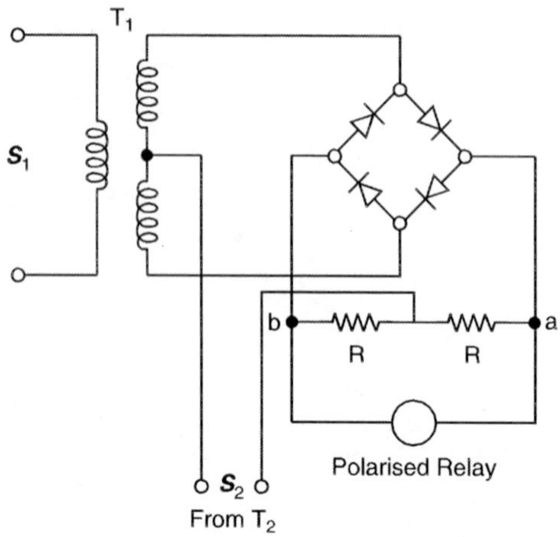

Fig 3.7 Rectifier bridge type phase comparator

compare the phases of S_1 and S_2, the rectifier bridge uses the principle of duality i.e., it compares the amplitudes of $(S_1 + S_2)$ and $(S_1 - S_2)$. The current I_{PR} flowing through the polarized relay is

$$I_{PR} = \left[|S_1 + S_2| - |S_1 - S_2| \right]$$

Thus the phases of S_1 and S_2 are continuously compared.

These type of comparators are widely used for the realization of distance relay characteristics.

3.4.3 Integrating type phase comparator

Figure 3.8 (a) shows the block diagram of an integrating type phase comparator. The AND gate output (coincident period of inputs) is fed to the integrator. The output of the integrator is applied to the level detector which gives the trip signal.

Figure 3.8 (b), (i) and (ii) shows the output of AND gate the integrator for $\phi < 90°$ i.e. operating condition. Figure 3.8 (b), (iii) and (iv) shows the output of AND gate and integrator for $\phi = 90°$ (i.e. limiting condition). Any value of coincidence period less than the limiting will not operate the trip circuit as shown in Fig 3.8 (b), (v) and (vi).

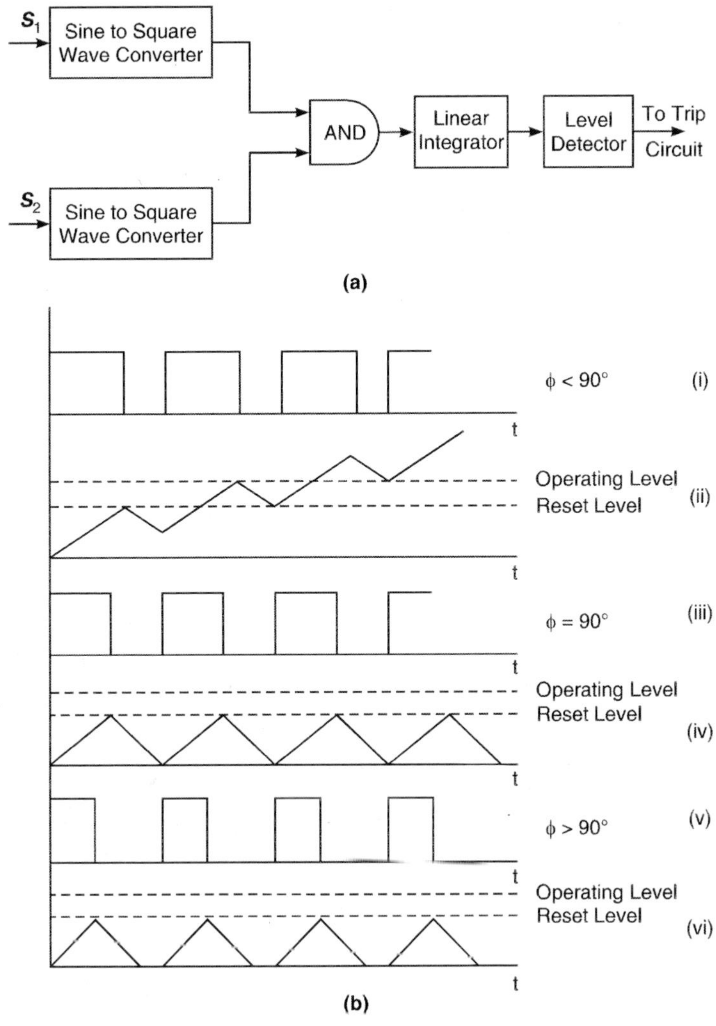

Fig 3.8 (a) Block diagram of an integrating type phase comparator
(b) Waveforms of an integrating type phase comparator

3.4.4 Phase Comparator Based on Pulsed Signals

The Signals S_1 and S_2 are converted to in-phase rectangular pulses $[S_1]$ and $[S_2]$ as shown in Fig.3.9 (a) and (b). The pulsed signals $[S_1]$ and $[S_2]$ (Fig 3.9 b) are added and subtracted to produce $[S_1]+[S_2]$ and $[S_1]-[S_2]$ as shown in figures 3.9 (c) and (d) respectively. The signals $[S_1]+[S_2]$ and $[S_1]-[S_2]$ are fed to a rectifier, averager and then to a phase comparator.

It can be seen from the Fig 3.9 that if $\phi < 90°$, the modulus of $[S_1]+[S_2]$ is greater than the modulus of $[S_1]-[S_2]$. That is

$$|\,[S_1]+[S_2]\,| > |\,[S_1]-[S_2]\,|\,; \quad -90° < \phi < 90°$$

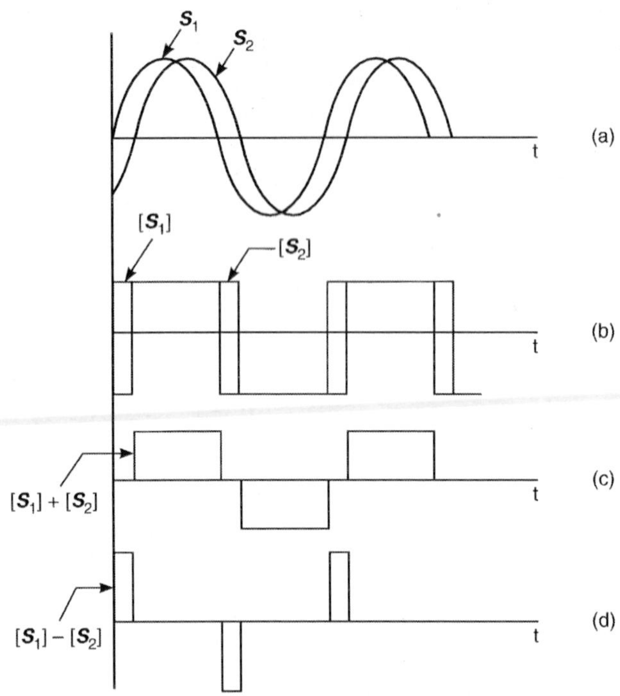

Fig 3.9 Phase comparison by pulsed signals ($\phi < 90°$)

If $\phi > 90°$

$$|\,[S_1]+[S_2]\,| < |\,[S_1]-[S_2]\,|\,; \quad 90° < \phi < -90°$$

The condition for $\phi > 90°$ are shown in Fig 3.10

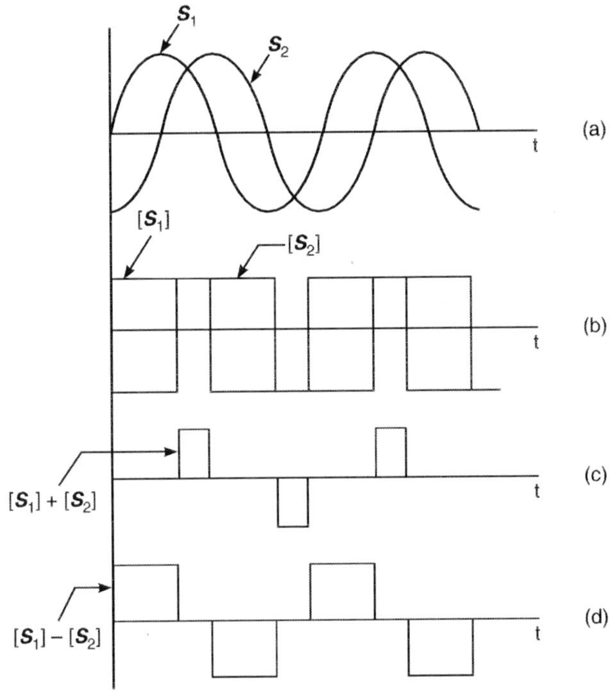

Fig 3.10 Phase comparison by pulsed signals ($\phi > 90°$)

3.5 DUALITY BETWEEN AMPLITUDE AND PHASE COMPARATORS

An amplitude comparator can be converted to a phase comparator and vice versa if the input quantities are altered to the original two input quantities.

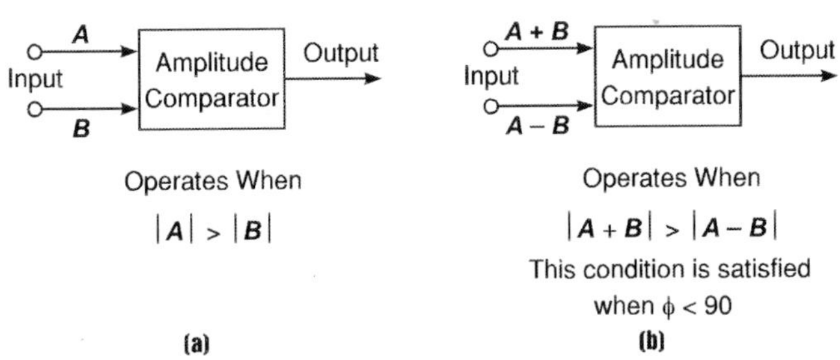

Fig 3.11 (a) Amplitude comparator
(b) Amplitude comparator used as phase comparator

Consider the operation of an amplitude comparator which has two phasor quantities A and B as input signals shown in Fig. 3.11(a). It operates when

$$|A| > |B|$$

Suppose that the input signals are changed to $(A + B)$ and $(A - B)$ as shown in Fig 3.11 (b). As the circuit is designed for amplitude comparison, now with changed input signals, it will operate when

$$|A + B| > |A - B|$$

Figure 3.12 shows the phasor diagrams for an amplitude comparator. It is seen that in Fig 3.12 (c) when the phase angle ϕ between A and B is less than 90°, the condition $|A + B| > |A - B|$ is satisfied and the relay operates.

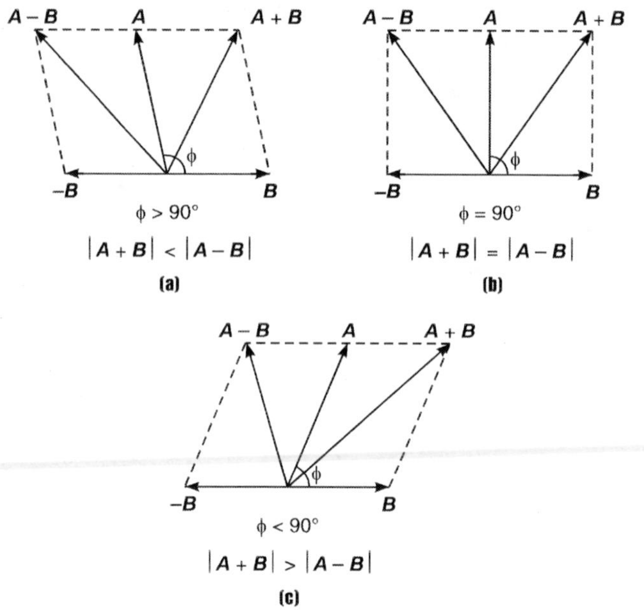

Fig 3.12. Phasor diagrams for amplitude comparator used for phase comparision.

When $\phi \geq 90°$ the relay does not operate.

Similarly, a phase comparator with inputs A and B [Fig 3.13(a)] operates when A and B have the same directional sense or polarity. Suppose that the input signals are changed to $(A + B)$ and $(A - B)$ as shown in Fig 3.13(b). Then the relay acts as an amplitude comparator when $|A + B| > |A - B|$ have the same directional sense, that is $|A| > |B|$. Under this condition the phase angle δ between $(A + B)$ and $(A - B)$ is less than 90°.

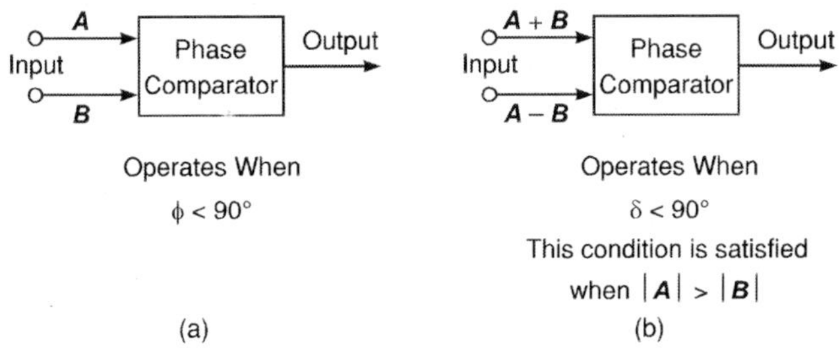

Fig 3.13. (a) Phase comparator
(b) Phase comparator used as amplitude comparator..

Figure 3.14 shows the phasor diagrams for a phase comparator. From Fig 3.14 (c) it is seen that δ becomes less than 90° only when $|\boldsymbol{A}| > |\boldsymbol{B}|$.

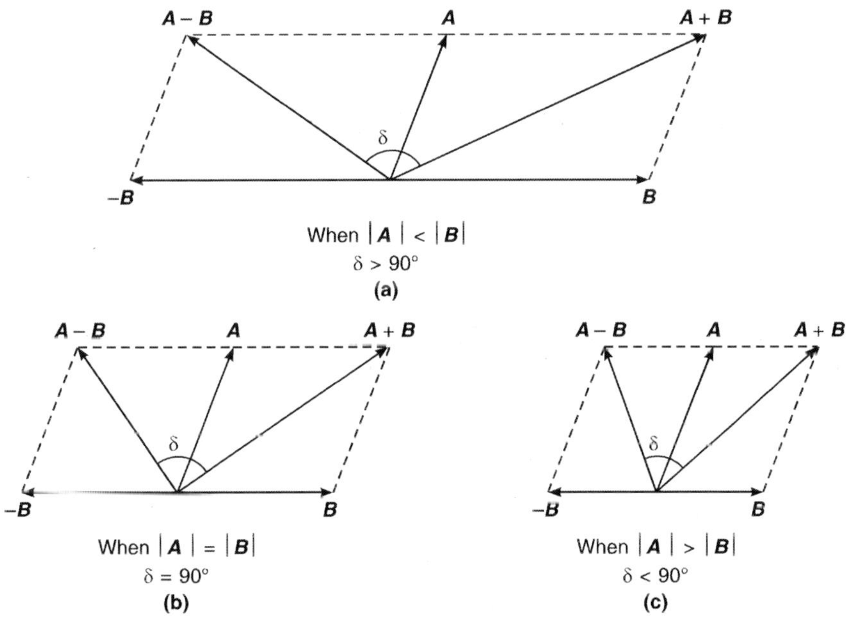

Fig 3.14. Phasor diagrams of phase comparator used for amplitude comparision

Therefore, an amplitude comparator can be changed to phase comparator and vice versa if the input quantities are changed to the sum and difference of the original two input quantities.

EXERCISES

1. What is a static relay ? Explain with the help of a block diagrams.
2. What are the advantages and limitations of static relays over electromagnetic relays.
3. Explain the concept of duality in static comparators.
4. What are the different types of amplitude comparators ? Describe the instantaneous type amplitude comparator
5. What is a phase comparator ? Describe pulsed signal based phased comparator.
6. Describe various types of phase comparators.
7. Discuss the coincidence principle used in phase comparators.
8. Discuss the operating principle of a rectifier bridge type comparator.
9. What are amplitude and phase comparators ? Name three types of each comparator.

❑❑❑

4

OVERCURRENT PROTECTION

4.1 INTRODUCTION

A protective relay which operates when the load current exceeds a preset value is called an *overcurrent relay*. The value of the preset current above which the relay operates is called its ***pick up current***. A scheme in which relays are used for protection of an element is called *overcurrent scheme* or *overcurrent protection*. Overcurrent protection is different than overload protection, which normally utilizes relays that operate in a time related to some degree to the thermal capability of the element to be protected. Whereas overcurrent protection is directed totally to the clearance of faults, even though with the settings usually adopted some degree of overload protection is achieved. Therefore, the maximum load currents must be known to determine whether the ratio of minimum fault current to maximum load current is high enough for the successful operation of simple overcurrent relays. Overcurrent relays are used for protection of transmission and distribution lines, large motors, power equipment, etc. Induction disc type electromagnetic relays are commonly used for overvoltage protection.

4.2 TIME-CURRENT CHARACTERISTICS

There are two basic forms of overcurrent relays, namely the instantaneous type and the time-delay type.

The instantaneous overcurrent relay is designed to operate with no intentional time delay when the current exceeds the relay settings. The time of operation of such relays is anywhere between 0.016 and 0.1 s.

The time-overcurrent relay has an operating characteristic such that its operation time changes inversely as the current flowing in the relay. The most important time-current characteristics used in overcurrent relays are given below. The name of the relay is according to its time-current characteristic.

Instantaneous Overcurrent Relay

Instantaneous overcurrent relay operates without any intentional time delay. The operating time is constant, irrespective of the magnitude of current as shown by curve (a) of Fig. 4.1. Its operating time is 0.15 or less. The terms ***high set*** or ***high speed*** overcurrent relays generally referred to instantaneous overcurrent relays.

Definite-time Overcurrent Relay

A definite-time overcurrent relay operates after a predetermined time when the current exceeds its pickup value. The operating time is constant, irrespective of the magnitude of current above the pickup value. A definite-time curve would strictly be one in which the operating time was unaffected by the magnitude of the actuating quantity.

Inverse-time Overcurrent Relay

An inverse-time overcurrent relay operates when the current exceeds its pickup value. The operating time depends upon the magnitude of the operating current. The operating time decreases as the current increases. Curve (b) of Fig. 4.1 shows the inverse-time current characteristic of this type of relay.

Inverse Definite Minimum Time (IDMT) Overcurrent Relay

This type of relay gives an inverse time characteristic at lower values of the fault current and definite time characteristic at higher values of the fault current as shown in Fig. 4.1 (c).

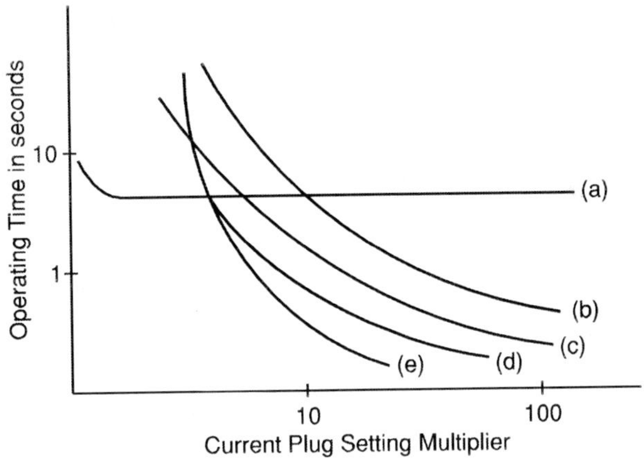

Fig. 4.1 Characteristics of various overcurrent relays (a) Instantaneous, (b) Inverse time, (c) IDMT, (d) Very inverse, and (e) Extremely inverse.

Very Inverse Relay

A very inverse relay gives more inverse characteristic than an IDMT relay. Very inverse characteristic is used to provide good selectivity as in case of ground faults and faults at the far end of the line.The characteristic is shown in Fig 4.1(d).

Extremely Inverse Relay

Extremely inverse relay provide even more inverse characteristic than a very inverse relay characteristic. Extremely inverse relays are employed where very inverse relays fail in selectivity. For example after a long outage a distribution system, when the system is closed the very inverse relay fails to distinguish between the rapid decaying inrush current of the load and persistent high current of a fault. The extremely inverse relay easily distinguishes between these two conditions due to steep time-current characteristic. An extremely inverse relay is also used for protection of machines against overheating as its characteristics ($I^2 t = k$) match with heating characteristics of the apparatus (heating is proportional to the square of the current). Hence, these types of relays are used for protection of machines and apparatus against overloading.

Sometimes even more steep characteristics are required for some special industrial applications. These characteristics can be realized by static or microprocessor based units.

Realization of Various Time-current Characteristics

For realization of different characteristics of electromagnetic relays, the magnetic circuit can be designed to saturate at different values of actuating current. In order to obtain the definite time characteristic the core is designed to saturate at pick up value of the current. For obtaining an IDMT characteristic the core made to saturate at a later stage. If the core is allowed to saturate at still later stage, very inverse characteristic stage is obtained. Even very later stage of saturation gives an extremely inverse characteristic.

Electromagnetic relays do not give straight line time-current characteristics. The general equation for inverse time current characteristic i.e. $t = \frac{k}{I^n}$ can give a straight line on $\log t - \log I$ graph for any value of n. For this purpose microprocessor based relays are used. The advantage of plotting these curves on log-log sheets is that simplified characteristics is obtained and saves considerable time in relay setting calculations.

4.3 SETTING OF INVERSE, TIME-DELAY OVERCURRENT RELAYS

There are two settings that must be applied to all time-delay overcurrent relays :

1. Pickup setting or current tap setting (CTS)
2. Time-delay setting (TDS)

4.3.1 Pickup Setting or Current Tap Setting

The relay operating characteristic is plotted with current (in amperes) on horizontal axis and operating time (in seconds) on vertical axis. It is seen to be an inverse-time relationship. The pickup of a relay is the minimum value of the operating current, voltage or other input quantity reached by progressive increases of the operating parameter that will cause the relay to reach its completely operated state when started from the reset condition. For example, an overcurrent relay set for 5A pickup will operate only if the current through the relay is 5A or greater. The setting is first calculated considering the maximum load and minimum fault currents in terms of primary current. Then the secondary current is calculated using the CT ratio. It is then specified in terms of the pickup of the relay being used so the relay can be set and calibrated.

In an induction type relay, there is a separate curve for each setting of the relay.

If the value of current is taken as a multiple of pickup current which is called the ***plug setting multiplier*** (PSM), the same curve can be used for any value of pick up current. That is, if the curves are known for a pickup value then the same curves can be used for all other multiples of that pickup value.

This type of pickup adjustment is possible in induction type relays in which the number of turns of the coil can be adjusted in order to have desired current settings. This is possible because the ampere-turns at pickup are the same for each tap. Hence at a given multiple of pickup, the coil ampere-turns and consequently the torque remains the same regardless of the tap used. These curves are plotted on log scales as shown in Fig 4.2.

PSM can be expressed as

$$\text{PSM} = \frac{\text{secondary current}}{\text{relay current setting}}$$

$$= \frac{\text{primary current during fault, (that is fault current)}}{\text{relay current setting} \times \text{C.T. ratio}}$$

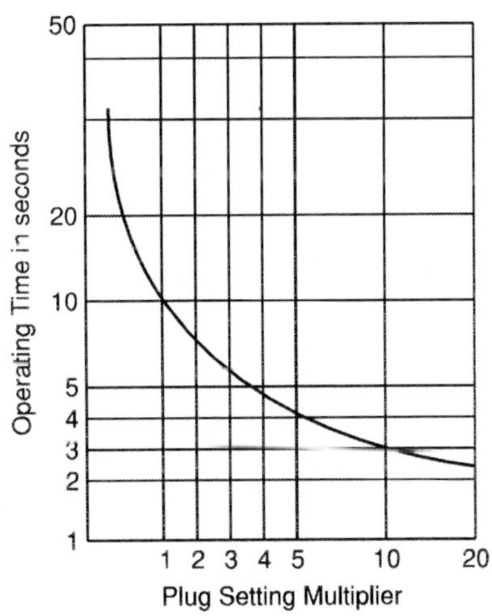

Fig. 4.2 Time-current characteristics

Example 4.1. A relay operates with a rated current of 5A has a relay setting of 150 percent. It is connected to a power system through a CT of 400/5 ratio. If the fault current is 6000A, determine the plug setting multiplier of the relay.

Solution : The operating current of the relay

$$= \text{pickup current} \times \text{relay setting}$$

$$= 5 \times \frac{150}{100} = 7.5 \text{ A}$$

$$\text{C.T. ratio} = \frac{400}{5} = 80$$

Plug setting multiplier of the relay

$$\text{PSM} = \frac{\text{fault current}}{\text{operating current of the relay} \times \text{C.T. ratio}}$$

$$= \frac{6000}{7.5 \times 80} = 10$$

4.3.2 Time-delay Setting

The time-delay feature of the relay is obtained in a variety of ways, depending on the design of the relay.

In an induction disc type relay the time of operation can be set at a desired value. This is done by adjusting the angular distance travelled by the disc of the relay to close the contacts. The angular distance travelled is divided into 10 equal steps say (0.1, 0.2, 0.3,..., 1.0). These steps of time settings are known as ***time multiplier setting (TMS)***.

The operating time can be calculated by multiplying the TMS = 1 for a particular plug setting.

In solid-state relays time-delay is obtained by using electrical R–L–C circuits.

The purpose of time-delay setting is to enable relays to coordinate with each other.

Example 4.2. An IDMT type overcurrent relay is used to protect a feeder through a 500/1 A CT. The relay has a PS of 125% and TMS = 0.3. Find the time of operation of the relay if a fault current of 5000 A flows through the feeder. The relay-characteristic curve is as follows :

PSM	2	4	5	8	10	15
Time for unity TMS (100% current = 1A)	10	6	4.5	3.2	3	2.5

Solution : Relay current $I_R = \frac{\text{fault current}}{\text{C.T. ratio}}$

$$= \frac{5000}{500/1} = 10 \text{ A}$$

Pickup current of relay

$$= \text{Current setting} \times \text{rated secondary current of CT}$$

$$= \frac{125}{100} \times 1 = 1.25 \text{ A}$$

Plug setting multiplier of the relay

$$\text{PSM} = \frac{\text{fault current in relay}}{\text{pickup current}}$$

$$= \frac{10}{1.25} = 8$$

Time corresponding to the PSM of 8 from the given characteristic curve is 3.2 s.

Actual operating time = PSM × TMS

$$= 3.2 \times 0.3 = 0.96 \text{ s}$$

4.4 RELAY COORDINATION

For transmission line protection in interconnected systems, it is necessary to provide the desired selectivity such that the relay operation results in the least service interruption while isolating the fault. This procedure is known as ***relay coordination***.

The following criteria are to be satisfied for relay coordination:

1. As far as possible, relays with the same operating characteristics are to be used in series with each other.
2. The relay farthest from the source must have the current settings equal to less than the relays behind it.
3. The relay settings are first selected to provide the shortest operating times at maximum fault current levels and then checked to determine whether or not they are satisfactory at minimum fault current levels.

There are various possible methods to achieve correct relay coordination. Time/current gradings are involved in three basic methods for radial or loop circuits where there are several line sections in series.

The three methods of relay grading are

(a) Time grading

(b) Current grading

(c) Combination of time and current grading.

4.4.1 Time Grading

Figure 4.3 shows a simple radial system. In the time grading method the selectivity is obtained on the basis of time of operation of the relays. The time of operation of the relays at various locations is so adjusted that the relay farthest from the source will have minimum time of operation. The time settings increase as the relay gets closer to the source.

In Fig. 4.3, a protection unit comprising a definite time-delay over current relay is placed at each of the points (2), (3) and (4). The

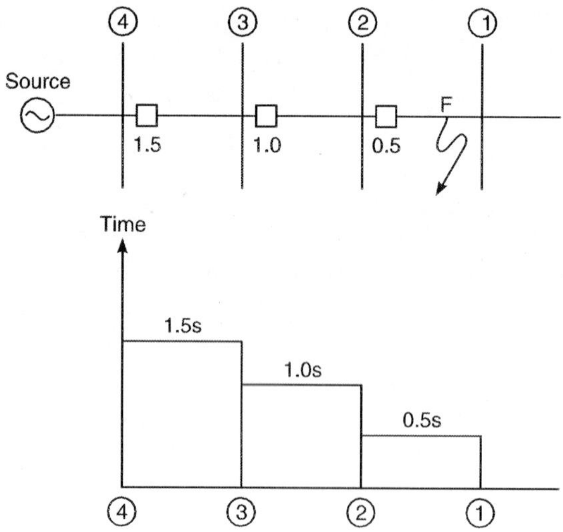

Fig 4.3 Time graded overcurrent protection of a feeder

Time delay of the relay provides the selectivity. The relay at the circuit breaker (2) is set at the shortest possible time necessary for the circuit breaker to operate (typically 0.5 second). The relay setting at (3) is chosen here as 1 second, and the relay setting at (4) is 1.5 second.

When a fault occurs at point F, the relays at (2), (3) and (4) come into action as the fault current flows through all of them. The relay at (2) will operate since it has the least time setting. Therefore, the circuit breaker at (2) will trip and the fault will be isolated. Now the relays at (3) and (4) are reset. If the relay or circuit breaker at (2) fails, the fault remains uncleared. Now after 1 second, the relay at (3) will operate and the circuit breaker at (3) will trip. If the relay or circuit breaker at (3) also fails to operate, then after 1.5 seconds, the relay at (4) will operate to trip the circuit breaker at (4).

The main drawback of this method is that the longest fault clearing time takes place for the faults in the section closest to the source where the faults are more severe.

4.4.2 Current Grading

The fault current along the length of the protected circuit decreases as the distance from the source to the fault location increases. This fact is utilized in the current grading method of achieving the selectivity of the relays. The relays controlling the various circuit breakers are set to operate at a suitably graded current setting that decreases as the distance from the source is increased. Figure 4.4 shows an example of a radial system with current grading at properly tapered values such that only the relay nearest to the fault trips the circuit breaker.

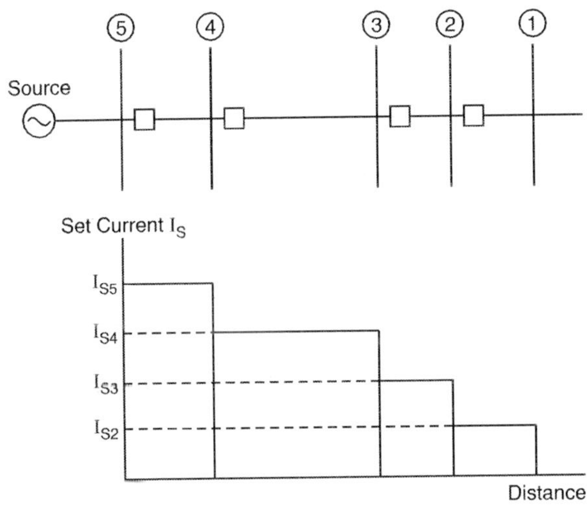

Fig. 4.4 Current grading for a radial system

In this method high set (high speed) instantaneous overcurrent relays are used. In practice, the magnitude of the fault current cannot be accurately determined as all the circuit parameters may not be known. The performance of instantaneous relays is not accurate under transient conditions. Furthermore the relay cannot differentiate between faults that are very close to each other (for example, on each side of a bus). This is due to the fact that the difference in the current would be extremely small.

4.4.3 Combination of Current and Time Grading

Either the time grading or the current grading alone for achieving the selectivity of relays has its limitations. Therefore, a combination of current and time grading is usually used. This method is known as ***time current grading***. IDMT overcurrent relays are used in this method. They have the combined features of time and current grading. IDMT relays have both time and current setting arrangements. The current setting of the relay is made according to the fault current level of the particular section of the feeder to be protected. The relays are set to pickup progressively at higher current levels towards the source. Note that the pickup current is the current that causes the relay to operate and close contacts. The pickup current is set by current tap settings. Time setting is also done in a progressively increasing order towards the source. The difference in operating times of two adjacent relays is usually kept 0.5 second.

The time-current grading method is widely used for the protection of distribution lines.

Overcurrent relaying is the simplest and cheapest, but it is most difficult to apply. Actually, there is no way to coordinate over current relays in a complex network. Further, with the growth of load or with the modification of the system configuration, the system fault levels at relay locations rise. The relays would then require resetting or even replacement. Therefore overcurrent relays cannot be applied. Instead distance relays would be employed. The distance relays are sensitive to load growth and configuration change.

Coordination of overcurrent relays is achieved by adjustment of pickup current and time inverseness of their characteristics. In the application of overcurrent relays, the CT ratio for each relay is selected and each relay is adjusted for its

- pickup current [plug setting multiplier (PSM)]
- time multiplier setting (TMS)

The current transformer (CT) used for each relay must be so selected that under full-load conditions, the relay carries current less than its rated continuous current (5A or 1A). This is assured if the relay current for maximum fault current is close to the rated current.

Disadvantages of Graded Overcurrent Systems

There are two disadvantages in the application of graded over current systems to feeder protection.

1. The grading settings may lead to tripping times that are too long to prevent damage and service interruptions.
2. Satisfactory grading for complex networks is quite difficult to attain.

4.4.4 Coordination Delay Time (CDT)

The time interval that is necessary between two adjacent relays is called the coordination delay time (CDT). It is the minimum interval that permits a relay and its circuit breaker to clear a fault in its operating zone. The CDT depends upon the following factors :

1. The fault current interrupting time of the circuit breaker. It is about 0.1 s or 5 cycles for the modern circuit breaker.
2. The overtravel of the relay (when the relay is deenergized, operation may continue for a little bit longer until stored energy is dissipated).
3. Errors due to relay and CT tolerances.
4. Errors due to calculated approximations.
5. Extra safety margin.

The sum of factors 2–5 is called **error margin**. It is of the order of 0.3s.

4.5 APPLICATION OF INVERSE, TIME-DELAY OVERCURRENT RELAYS TO A SERIES OF RADIAL LINES

The principle of relay coordination can be explained by reference to Fig. 4.5 which shows a series of radial lines and the time-distance characteristics of the associated inverse time-delay overcurrent relays. The curves are plotted as a function of fault location. Since the magnitude of the fault current decreases as the fault moves away from the source,these curves appear 'reversed' when compared to the time-delay overcurrent relay operating characteristics. Figure 4.5 shows how time coordination is achieved between inverse time overcurrent relays at each circuit breaker location. In order to have a proper coordination between various relays on a radial system, the relay farthest from the source must be set to operate in the minimum possible time. The time settings are increased for the succeeding relays towards the source. For inverse-time overcurrent relays, the time setting is determined based on the maximum fault current. If the relay has correct discrimination at maximum fault current, it will automatically have a greater discrimination at the minimum

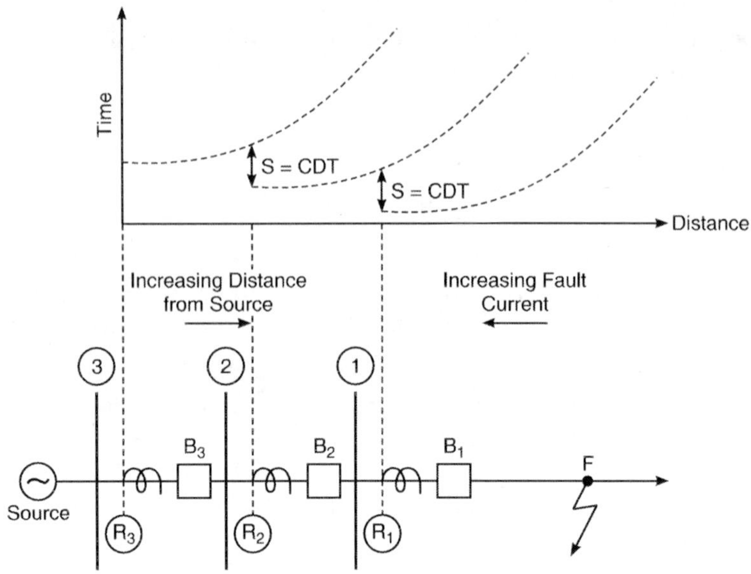

Fig. 4.5 Relay coordination principles

fault current because the characteristic curve is more inverse on a lower current region.

A vertical line drawn through any assumed fault location will intersect the operating time curves of various relays and will thereby show the time at which each relay would operate if the fault current continued to flow for that length of time.

For fault F at the end farthest from the source, relay R_1 tripling breaker B_1 operates first at time T_1. Relay R_2 tripping breaker T_2 operates next. The operating time T_2 of relay R_2 is given by

$$T_2 = T_1 + \text{CDT}$$

where CDT = (operatign time of breaker B_1) + error margin.

Similarly, relay R_3 at breaker B_3 coordinates with the relay at breaker B_2 by having a still longer time delay (including the same coordinating time (S). The operating time T_3 of the relay at bus 3 can be expressed as

$$T_3 = T_2 + \text{CDT}$$

It is to be noted that R_3 at breaker 3 has the longest time delay and will not trip unless none of the other breakers trips, provided that the fault current is greater than its pickup setting.

4.6 GENERAL RELAY EQUATION

The general relation equation for the operation of overcurrent, directional and impedance relays is given by

$$Q = K_1 I^2 + K_2 V^2 + K_3 VI \cos(\phi - \alpha) - K_4 \quad \text{...(4.6.1)}$$

where,

V = rms magnitude of voltage

I = rms magnitude of current

ϕ = phase angle between **V** and **I** phasors

α = an adjustable parameter

The relay operates when

$$Q > 0 \quad \text{...(4.6.2)}$$

In the general relay equation Q is the torque in an electromagnetic relay and it is the voltage or current in a solid-state relay.

4.7 DIRECTIONAL RELAYS

The general torque equation of an electromagnetic relay is given by

$$\tau = K_1 I^2 + K_2 V^2 + K_3 VI \cos(\phi - \alpha) - K_4 \quad \text{...(4.7.1)}$$

If $K_1 = 0,\ K_2 = 0,\ K_4 = 0$

then $$\tau = K_3 VI \cos(\phi - \alpha) \quad \text{...(4.7.2)}$$

Equation (4.6.2) represents the torque relationship of a directional relay. Thus a directional relay is a two input device. In Eq (4.7.2)

V = VT secondary voltage proportional to system line-to-neutral voltage

I = CT secondary current proportional to system line current

ϕ = phase angle by which **I** lags **V**

The relay operates when

$$\tau > 0$$

$$K_3 VI \cos(\phi - \alpha) > 0$$

Since $K_3\ V, I$ are not zero therefore

$$\cos(\phi - \alpha) > 0$$

For $\tau < 0$, that is, $\cos(\phi - \alpha) < 0$

the relay blocks.

The parameter α in the Eq (4.7.2) adjustable and is taken to be equal to 90°. Therefore Eq (4.7.2) becomes

$$\tau = K_3 \, VI \cos(\phi - 90°)$$

or $$= K_3 \, VI \sin \phi \qquad ...(4.7.3)$$

It is seen from relation (4.7.3) that an electromagnetic directional relay senses sine of the phase angle ϕ. Hence an electromagnetic relay is a sine phase comparator.

In a directional relay, V is taken as a reference phasor (polarizing quantity). In case of a fault there is only a slight change in the phase angle of V while the angle of the fault current can change by as large as 180° depending on whether the fault is within or outside the protected zone. Thus a directional relay is able to discriminate a fault on the protected zone or that outside it.

4.8 REVERSE POWER OR DIRECTIONAL RELAY

Figure 4.6 shows an electromagnetic directional relay. It consists of two electromagnets. The upper electromagnet has two windings, namely, the primary winding and the secondary winding. The primary is connected to the secondary of the CT in the line to be protected and is provided with tappings. The tappings are connected to a plug setting bridge by which the number of turns in use can be adjusted. Thus, the desired current setting can be obtained. The secondary winding and the winding of the lower electromagnet are in series and are energized by induction. A thin nonmagnetic metallic disc is mounted on a spindle between the two electromagnets so that it cuts the fluxes of both the magnets. A deflecting torque is produced by the flux of each magnet with the eddy current induced in the disc by the flux of the other magnet. The torque is proportional to VI cos ϕ, where ϕ is the phase angle between V and I. The torque is maximum when voltage and current are in phase. During fault condition, the power factor is very poor. To produce maximum torque during the fault conditions, a compensating winding and copper shading band are used as shown in Fig. 4.6. The spindle of this disc carries a moving contact which bridges the two fixed contacts when the disc rotates through an angle which is adjustable between 0° to 360°. By adjusting this angle the travel of the moving contact can be adjusted so that the relay can be provided any desired time setting which is indicated on a time setting dial.

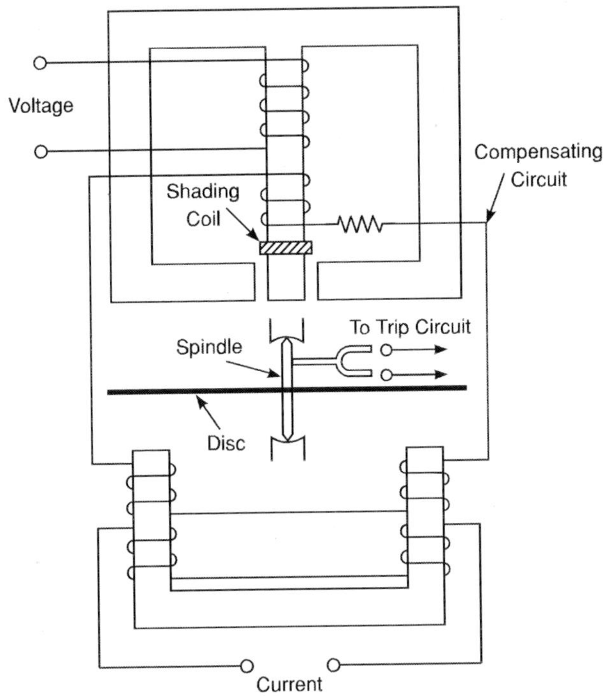

Fig. 4.6 Induction type directional relay

4.9 INDUCTION TYPE DIRECTIONAL OVERCURRENT RELAY

Figure 4.7 shows an induction type directional overcurrent relay. It consists of a combination of an overcurrent unit and a directional unit. The secondary winding of the over current unit is open when the power flow is in the normal direction. When the direction of power flow reverses, the directional unit operates. It is to be noted that overcurrent unit is not energized unless the directional unit contacts are closed because the operating coil of the overcurrent unit completes it circuit through the directional unit contacts. Thus the overcurrent unit operates only after the directional unit contacts close. Now, if there is an overcurrent, the contacts of the overcurrent unit close to trip the circuit breaker.

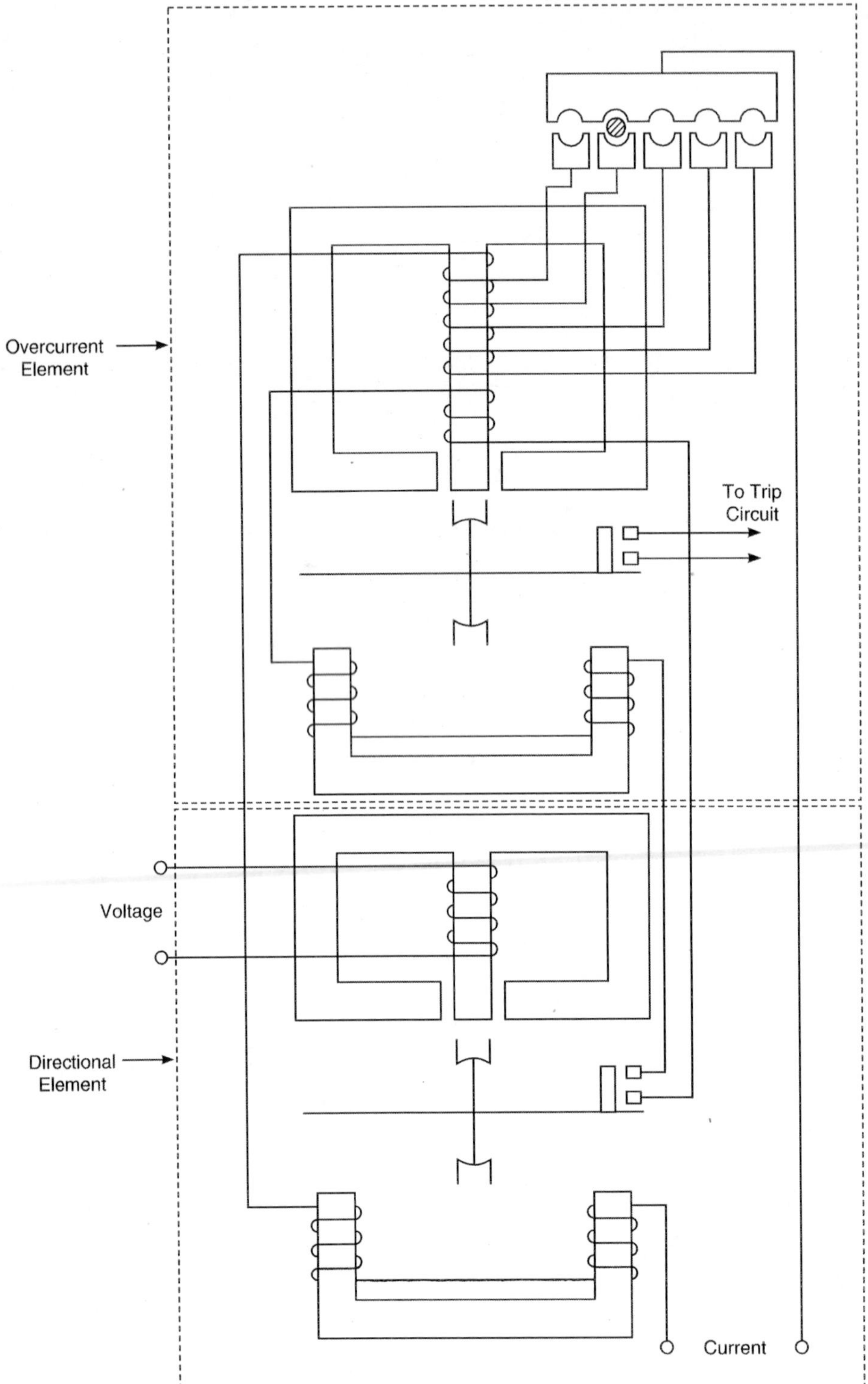

Fig. 4.7 Induction type directional overcurrent relay

4.10 APPLICATION OF OVERCURRENT RELAYS

Overcurrent relaying is the simplest and cheapest. The principal application of overcurrent relays is on a radial system where they provide both phase and ground protection. Overcurrent relays are also used on industrial systems and on subtransmission lines. Even though the application of overcurrent relaying on radial circuits is relatively simple, its application on the loop and/or interconnected circuits becomes the most difficult since it requires readjustment as the system configuration changes. Overcurrent relays cannot discriminate between load and fault current, and therefore, when they are used for phase fault protection, they are only applicable when the minimum fault current is larger than the maximum full-load current.

4.11 APPLICATION OF DIRECTIONAL OVERCURRENT RELAYS

Directional overcurrent relaying is necessary for multiple source circuits, when it is essential to limit relay tripping for faults in only one direction. It is impossible to obtain correct relay selectivity by the use of non-directional overcurrent relay in such cases. If the same magnitude of the fault current could flow in either direction at the relay location, coordination with relays in front of, and behind, the non-directional relay cannot be achieved. Therefore, overcurrent relaying is made directional to provide relay coordination between all the relays that can see a given fault.

4.12 PROTECTION OF PARALLEL FEEDERS

To illustrate the need of directionality, consider single-end-fed system of parallel feeders shown in Fig. 4.8. At the sending end of the feeders (at 1 and 3), non-directional relays are required. The symbol ↔ indicates a non-directional relay. At the other end of feeders (at locations 2 and 4),

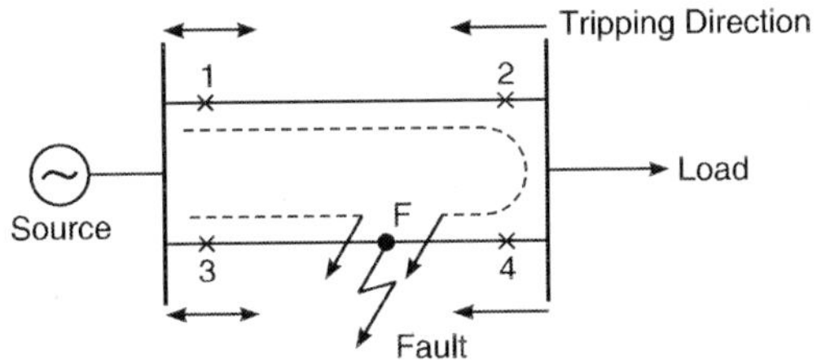

Fig. 4.8 Parallel feeders, single-end fed system.

directional overcurrent relays are required. The arrow mark → for directional relays located at 2 and 4 indicate that the relay will operate if the current flows in the direction shown by the arrow. Suppose that a fault occurs at *F*. The flow of fault current is shown by dotted path. The direction of the fault current is reversed in the directional relay 4 and, therefore, the relay at 4 trips. Since the fault current in the directional relay at 2 is in the normal direction, it does not trip. The relay located at 3 trips for a fault at *F*. Thus, the faulty feeder is isolated and the supply of the healthy feeder is maintained.

If non-directional relays are used at 2 and 4, both relays located at 2 and 4 will trip for a fault at *F*. This will result in both lines being tripped out for a fault at *F*. This is not desirable. Therefore, directional overcurrent relays are used at 2 and 4. Since the direction of current at 1 and 3 does not change for a fault on any of the parallel lines, non-directional relays are used at 1 and 3. Thus, in a single-end-fed system of parallel feeders directional overcurrent relays are provided in conjunction with nondirectional overcurrent relays.

It is to be noted that all directional relays have their tripping direction away from the concerned bus.

4.13 PROVISION OF DIRECTIONALITY TO AN OVERCURRENT RELAY

There are two approaches to providing directionality to an overcurrent relay :

1. Directional control
2. Directional overcurrent

Directional Control

In this type of control, the design of the relay is such that the overcurrent element will not operate until the directional element operates, indicating that the fault is in the tripping direction. This is shown in Fig. 4.9a.

In electromechanical relays this is done by using the directional element contact in series with the overcurrent element coil. Therefore, no torque can be developed until the directional element contact is closed. In solid-state and digital relays this is done by logic circuitry or the algorithm.

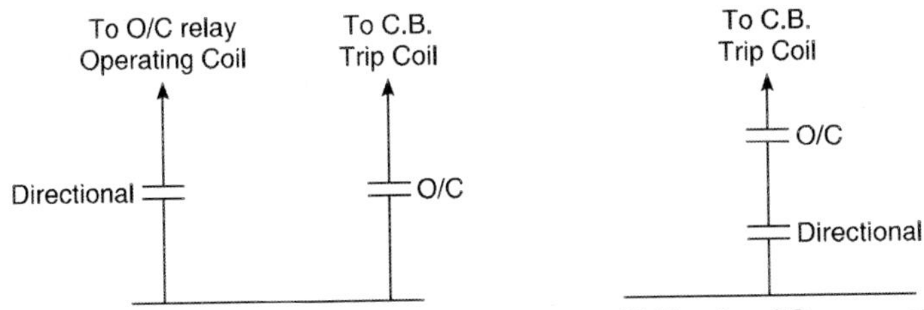

(a) Directional Control (b) Directional Overcurrent

Fig. 4.9 Directional control and directional overcurrent.

Directional Overcurrent

In this scheme, this relay has independent contacts, connected in series with the circuit breaker trip coil. Both relay contacts must close before a trip output is obtained. Directional overcurrent is shown in Fig. 4.9b.

EXERCISES

1. Explain in case of overcurrent relays
   (i) Instant time current characteristics
   (ii) Definite time current characteristics
   (iii) Inverse current characteristics
   (iv) Time current characteristics
2. What is an IDMT characteristic ? How it can be achieved in practice ?
3. Why IDMT characteristic is preferred to simple inverse characteristic.
4. Describe the techniques used to realize various time-current characteristics using electromagnetic relays.
5. Explain current time characteristics of the relays ans setting of relays for different applications.
6. Classify various type of overcurrent relays and give their application alongwith approximate characteristics.
7. What are various overcurrent protective schemes. Discuss their merits, demerits and fields of application.
8. What are the functions of current and time multiplier settings associated with such a relay.
9. What do you mean by time multiplier setting and plug multiplier setting of an overcurrent relay ? Explain with the help of relay characteristics.

10. What is meant by directional feature of a directional overcurrent relay ? Describe the construction, principle of operation and application of a directional overcurrent relay.

11. Explain with a neat sketch the operation of an induction type overcurrent relay

12. Explain the principle of working of a non directional induction type relay.

13. Show that the torque on the disc of an induction disc relay is maximum when the phase difference between two fluxes is 90°. Indicate the direction of rotation of the disc with reference to fluxes under the poles.

❑❑❑

5

DISTANCE PROTECTION

5.1 REACH OF RELAYS

The term ***reach*** is commonly used in connection with distance relays. A distance relay operates when the impedance (or a component of the impedance) as seen by the relay is less than a preset value. This preset impedance (or a component of impedance) or corresponding distance is called the reach of the relay. Thus, reach is the maximum length of the line up to which the relay can protect. A reach of 100% means that the relay is set to detect faults at any point up to the far end of the protected line.

Overreach means that the relay operates even when a fault point is beyond its preset reach (that is, its protected length).

Sometimes a relay may fail to operate even when the fault point is within its reach, but it is at the far end of the protected line. This is known as ***underreach*** of the relay.

Overreach may occur due to transients during the fault condition.

Underreach may be caused due to arc resistance, intermediate current sources, errors in C.T., V.T., errors in data available for initial setting of the relay and errors in measuring units.

5.2 THREE-ZONE PROTECTION

The transmission line having successive line sections can be protected by means of three-zone distance protection schemes. Consider a transmission line having three sections. The first section is AB, the second is BC and the third section is beyond C as shown in Fig 5.1. Three units of impedance relays are required at a particular location for three zones of protection. Usually, the first relay unit at A is typically set to protect only 80 percent of the line section AB. The protected zone of the first unit is called the ***first zone of protection***. The first relay unit is a high-speed unit and its operation is instantaneous. It is used for the primary protection of the line section AB. It is to be noted that the first relay unit is not set to protect the entire line section AB in order to avoid undesired tripping due to overreach.

The second relay unit at A is used to protect the remaining 20 percent of the line section AB which is beyond the reach of the first relay unit A. In order to take into account the effect of underreach, the second zone is extended into the adjoining line section AB. Usually, the setting relay unit is so adjusted that it protects the remaining 20 percent of the first line section AB and about 50 percent of the adjoining line section BC. The protective zone of the second relay unit at A is called the ***second zone of protection***. The second zone relay unit at A operates after a certain time delay. The operating time of the second zone is usually 0.2 s to 0.5 s.

The third zone relay unit at A is used to provide backup protection for faults in the adjoining line section BC. Its reach is extended beyond C under maximum underreach. The protective zone of the third stage is known as the ***third zone of protection***. The setting of the third relay unit at A covers the first line section AB, plus the second line section BC plus 25 percent of the third line section beyond C. The time delay for the third zone is usually 0.4 s to 1.0 s.

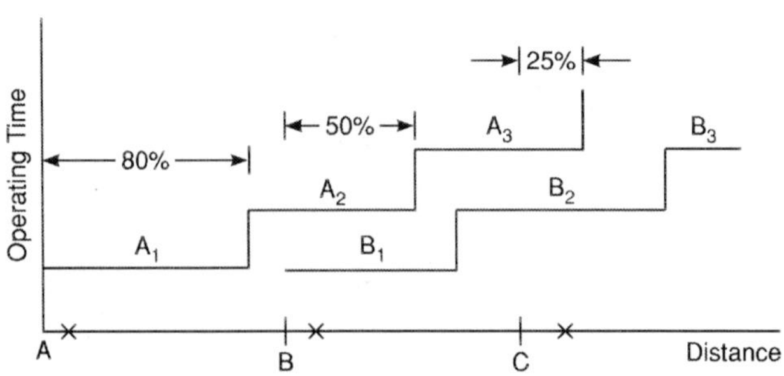

Fig. 5.1 Stepped time-distance characteristics of impedance relays

The idea of three-zone protection is to have each point in the system fall within at least two zones of protection, so there is a primary and backup protection everywhere.

The operating time characteristics of distance relays for a three-section line are shown in Fig.5.1. These are known as stepped time-distance characteristics. The operating times for the first, second and third zone relays placed at A are A_1, A_2 and A_3 respectively. Similarly, B_1, B_2, B_3 are operating times for first, second and third zone relays, (placed at B), respectively.

In order to save cost and panel space, it is not desirable to use three units of impedance relays at a particular location for three zones of protection. Present practice is to use only one measuring unit for all the three zones of protection. The ohmic reach is progressively increased by the timing unit to obtain the distance settings for the second and third zones.

5.3 OHMIC SETTING OF DISTANCE RELAYS

The ohmic setting of distance relays can be adjusted by altering the relationship of the ampere-turns of the operating coil to those of the restraining coil. Thus, it is made possible to select a setting suitable for the length of the line to be protected. The locus of points where the operating and resetting torques are equal is the ***threshold characteristic***. It depends on the ratio of voltage to current and the phase angle between them and is generally plotted on an R-X diagram. Distance relay (impedance relay) is basically a voltage restrained overcurrent relay.

5.3.1 R-X Diagram

In general, all electromechanical relays respond to one or more of the conventional torque producing quantities : (a) voltage, (b) current, (c) product of voltage and current and the angle θ between them and (d) a physical or design force such as control spring. Similar consideration hold for solid-state relays also. For a distance relay analyzing the response of the relay for all conditions is difficult because the voltage varies for each fault, or varies for the same fault but with different system conditions.

In order to overcome this difficulty, it is common to use an R-X diagram for analyzing and visualizing the relay response. By using two quantities R and X (or Z and θ), the confusion introduced by using three quantities E, 1 and θ is avoided. There is an additional significant advantage in that the R-X diagram allows to represent both the relay and the system on the same diagram.

5.4 DISTANCE RELAYS

The term ***distance relay*** is used for a family of relays that respond to the ratio of voltage to current and therefore to impedance or a component of impedance. A distance relay responds to input quantities as a function of electrical circuit distance between the relay location and the point of fault. Distance protection schemes are widely used for protection of transmission lines. The impedance of the transmission line is proportional to its length. Distance relay measures the impedance between the fault point and the relay location and operates for faults within its operating zone.

The distance relays are classified according to their polar characteristics, number of inputs they have, and the method by which comparison between the quantities is made. The common types compare two input quantities in either magnitude or phase to obtain characteristics that are either straight lines or circles on the R-X plane.

5.4.1 Types of Distance Relays

According to the shapes of the operating zones, the distance relays are classified as follows :

(a) Impedance relays

(b) Admittance or mho relays

(c) Reactance relays

(d) Quadrilateral relays

There four relay characteristic shapes are shown in Fig. 5.2.

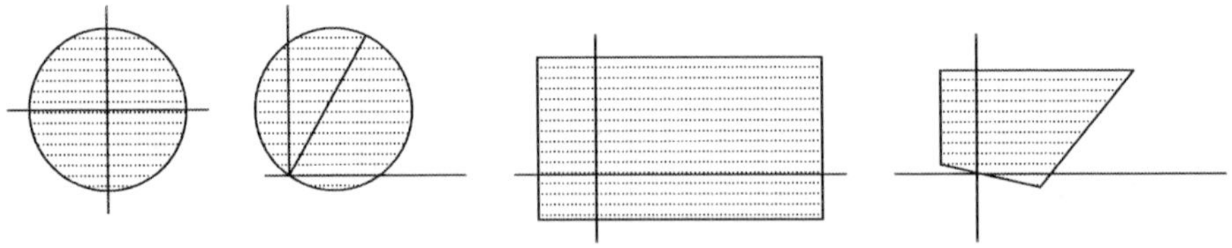

(a) (b) (c) (d)

Fig. 5.2 Types of distance relay characteristics : (a) Impedance relay (b) Admittance or mho relay (c) Reactance relay (d) Quadrilateral relay

The impedance relay has a circular shape which has its centre at the origin of the R-X diagram.

The admittance (or mho) relay has a circular shape which passes through the origin.

The reactance relay has a zone boundary defined by a line parallel to the R axis.

The quadrilateral characteristic is defined by four straight lines.

The quadrilateral characteristic is only available in solid-state or computer relays. More complex shapes can be obtained by using one or more of the above relay types, in a logical combination to provide a componsite tripping zone boundary.

The basic distance relays types are

1. Impedance relays
2. Reactance relays
3. Admittance or Mho relays
4. Offset mho or modified impedance relays
5. Ohm relays

5.5 IMPEDANCE RELAYS

An impedance relay measures the impedance of the line at the relay location. When a fault occurs on the protected line section, the measured impedance is the impedance of the line section between the relay location and the point of fault. The impedance of the line is proportional to its length, and hence, to the distance along the line. A distance relay is designed to operate only for faults taking place between the relay location and the point of fault.

In an impedance relay, current is compared with the voltage at the relay location. The current produce, a positive torque (operating torque) and the voltage produces a negative torque (restraining torque). The torque equation of an electromagnetic relay is given by

$$\tau = K_1I^2 - K_2V^2 - K_3 \qquad ...(5.5.1)$$

where K_1, K_2, K_3 are constants. The constant K_3 indicates spring control effect.

Since the spring effect is very small it may be neglected and $K_3 = 0$. Therefore, the torque equation can be written as

$$\tau = K_1I^2 - K_2V^2$$

The following condition should be satisfied for the operation of the relay

$$K_1I^2 > K_2V^2$$

or

$$K_2V^2 < K_1I^2$$

or $$\frac{V}{I} < \sqrt{\frac{K_1}{K_2}}$$

or $$\frac{V}{I} < K$$

or $$Z < K$$

where K is a constant. $\left(K = \sqrt{\frac{K_1}{K_2}}\right)$

For static relays and microprocessor based relays, I is compared with V. For operation of the relay, the following condition should be satisfied.

$$K_1 I > K_2 V$$

or $$K_2 V < K_1 I$$

or $$\frac{V}{I} < \frac{K_1}{K_2}$$

$$Z < K_a$$

where K_a is a constant $\left(K_a = \frac{K_1}{K_2}\right)$

These relations show that the relay is on the verge of operation when the ratio of V to I, that is, the measured value of the impedance is equal to a given constant. The relay operates when the measured impedance Z is less than the given constant.

The operating characteristic of an impedance relay in terms of voltage and current is shown in Fig. 5.3.

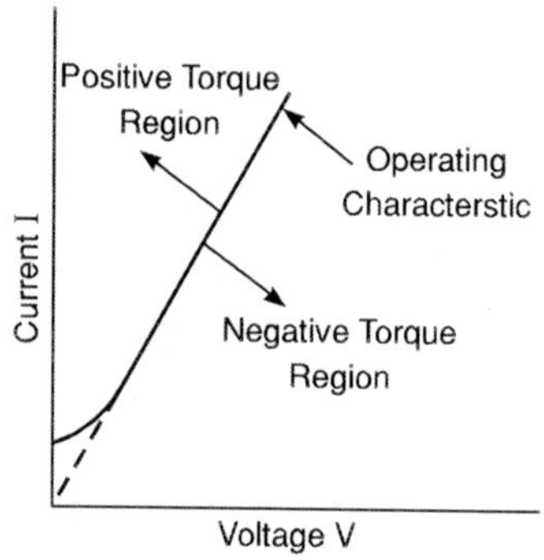

Fig. 5.3 Operating characteristics of an impedance relay

In an electromagnetic relay, the characteristic is slightly bent near the origin due to the effect of control spring. For a static relay microprocessor-based relays, the characteristic is a straight line.

The impedance relay does not consider the phase angle ϕ between the current and voltage applied to it. Therefore, its operating characteristic when plotted on the R-X diagram is a circle with its centre at the origin of the coordinates and of radius equal to $Z = K$ as shown in Fig. 5.4. The radius, Z is the setting of the relay and K is equal to the impedance of the line which is to be protected. The relay operates for all impedance values which are less than its setting, that is, for all points within the cross hatched circle. The relay restrains for all points outside the circle. In other words, the region within the circle is the operating zone and the region outside the circle is the blocking zone.

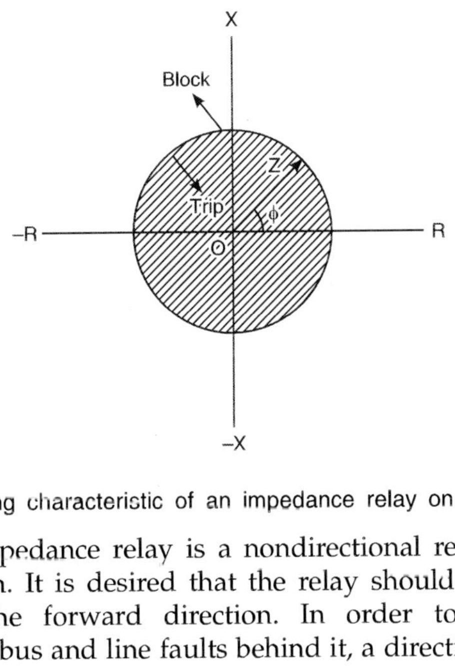

Fig. 5.4 Operating characteristic of an impedance relay on the R-X diagram

The plain impedance relay is a nondirectional relay. It can operate in either direction. It is desired that the relay should operate for faults lying only in the forward direction. In order to make the relay nonresponsive to bus and line faults behind it, a directional unit is added to this type of relay.

For a directional relay

$$Q = K_3\, VI \cos(\phi - \theta) - K_4$$

where ϕ is the phase angle between voltage phasor V and current phasor I and θ is the impedance angle of the relay. Values of angles are taken as positive in the clockwise direction.

For small K_4, the relay operates when

$$Q = K_3\, VI \cos(\phi - \theta) > 0$$

Since V, I and K_3 are not zero

$$\cos(\phi - \theta) > 0 \text{ for tripping}$$

and $$\cos(\phi - \theta) < 0 \text{ for blocking}$$

The impedance characteristic of the directional relay is a straight line on the R-X diagram.The combined characteristic of the directional and impedance relays is the cross hatched semi circle shown in Fig. 5.5.

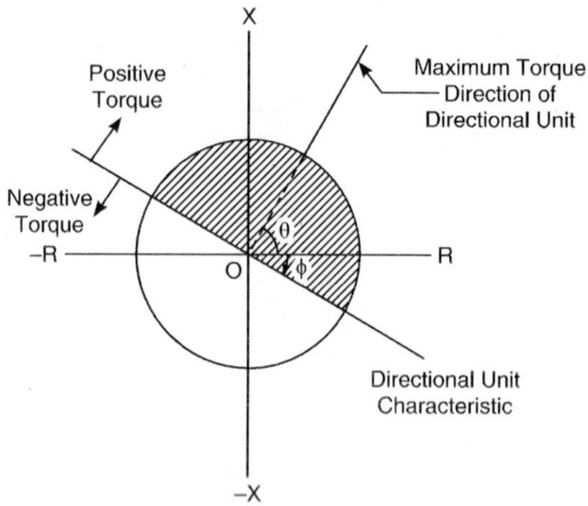

Fig. 5.5 Characteristics of impedance type distance relay with directional restraint.

Disadvantages of Impedance Relays

1. It is nondirectional. Therefore, it will see faults both in front of and behind the relaying point. Thus, it requires a directional unit to provide it with correct discrimination.
2. It is affected by arc resistance during arcing faults on the protected line.
3. It is highly sensitive to power swings, due to the large area covered by the impedance circle.

5.6 REACTANCE TYPE DISTANCE RELAY

The reactance-type distance relay consists of two units:

(i) an overcurrent element developing positive torque, and

(ii) a current-voltage directional element which either aids or opposes the overcurrent element, depending on the phase angle between the current and voltage. Hence, *a reactance relay is an overcurrent relay with directional constraint*. The directional element is arranged to develop maximum torque when the current lags its voltage by 90°.

If the control-spring torque is represented by $-K_3$, the torque equation is

$$\tau = K_1 I^2 - K_2 VI \sin\phi - K_3$$

where ϕ is defined positive when I lags V. At balanced point, the net torque is zero, and hence

$$K_2 VI \sin\phi = K_1 I^2 - K_3$$

Dividing both the side by $K_2 I^2$, we get

$$\frac{V}{I}\sin\phi = \frac{K_1}{K_2} - \frac{K_3}{K_2 I^2}$$

or

$$Z\sin\phi = \frac{K_1}{K_2} - \frac{K_3}{K_2 I^2}$$

But

$$Z\sin\phi = X$$

$$X = \frac{K_1}{K_2} - \frac{K_3}{K_2 I^2}$$

If the effect of control spring is neglected *i.e.*, $K_3 = 0$

$$X = \frac{K_1}{K_2} = \text{constant}$$

Thus, this relay has an operating characteristic such that for all positions of the impedance phasor Z, whose heads lie on this characteristic will have constant X component. That is, the operating characteristic of a reactance will be a straight line parallel to R axis as shown in Fig. 5.6.

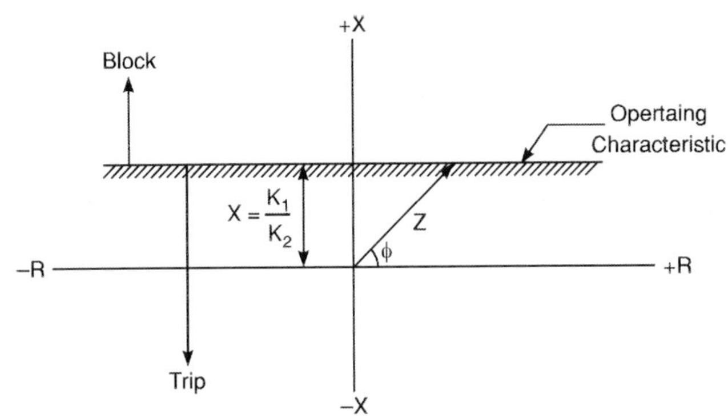

Fig. 5.6 Operating characteristic of a reactance-type relay

It is to be noted that the resistance component of the impedance has no effect on the operation of the relay. In case of a fault on the protected line, the measured reactance is the reactance of the line between the relay location and the fault point. The relay responds only to the reactance component. Thus, the relay does not sense resistance under fault conditions, that is it is insensitive to the resistance of an arcing type fault.

Any point below the operating characteristic - whether above or below the R axis - will lie in the positive-torque region (that is, the trip region).

A reactance relay is a non directional relay as it will also operate for the negative values of X. The negative value of X means that the fault is behind the relay location, that is, in the reverse direction.

The operation of a reactance relay is not affected by fault arc resistance since it responds only to the reactive component of the system impedance. However, when fault arc resistance is high such that the load and fault current magnitudes are of the same order the reach of the relay is altered by the value of the load and its power factor. Therefore, in order to provide directional feature and to prevent operation under normal load conditions, a voltage restraining unit (mho relay) is added to the relay. Such a modified reactance relay is called ***starting relay***.

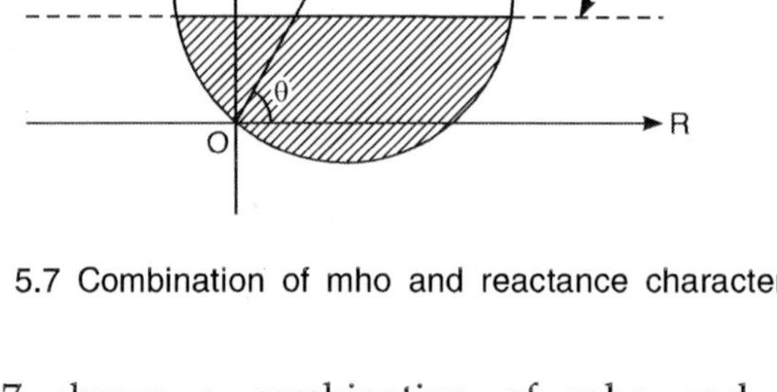

Fig. 5.7 Combination of mho and reactance characteristics

Figure 5.7 shows a combination of mho and reactance relay characteristic. Figure 5.8 shows the application of reactance for zone protection, In this figure, O is the relay location, OA is the first line section, AB is the second line section, BC is the third line section and θ is the line impedance angle.

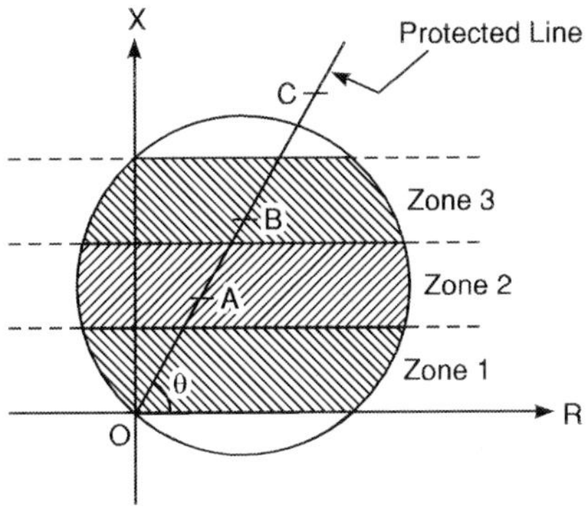

Fig 5.8 Application for zone protection

5.7 ADMITTANCE RELAY (MHO RELAY)

In an admittance relay or mho relay, the operating torque is obtained by the voltage-current directional element and restraining torque due to voltage element. In other words, *a mho relay is a voltage-restrained directional relay.*

The torque equation of a mho relay is given by

$$\tau = K_1\, VI \cos(\phi - \theta) - K_2\, V^2 - K_3$$

At balance point, $\tau = 0$

$$K_2 V^2 = K_1\, VI \cos(\phi - \theta) - K_3$$

Dividing both sides by $K_2\, VI$ we get

$$\frac{V}{I} = Z = \frac{K_1}{K_2} \cos(\phi - \theta) - \frac{K_3}{K_2 VI}$$

If we neglect the control spring effect, $K_3 = 0$ and

$$Z = \frac{K_1}{K_2} \cos(\phi - \theta)$$

$$Y \cos(\phi - \theta) = \frac{K_2}{K_1}$$

This is the equation of a circle of diameter $\frac{K_1}{K_2} = Z_R$ (say), the ohmic setting of the relay, which passes through the origin. Therefore, the operating characteristics of a mho relay when drawn on R–X diagram is a circle passing through the origin which makes the relay inherently directional. This characteristics is shown in 5.9(a).

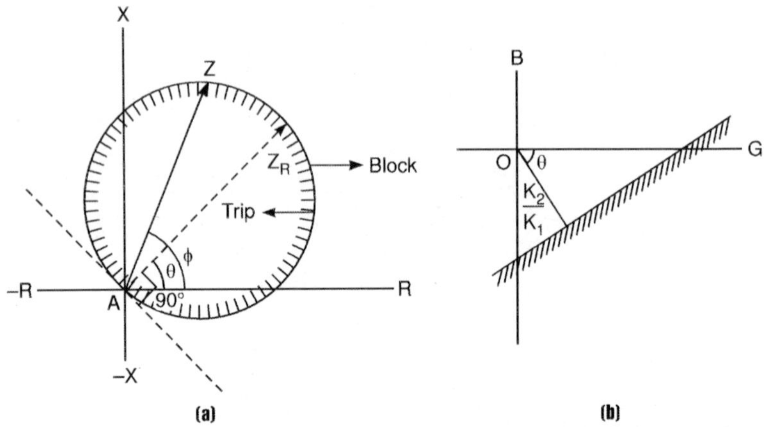

Fig. 5.9 (a) Operating characteristics of mho type relay on *R–X* plane
(b) Operating characteristics of mho type relay on *G–B* plane

It is also called an ***admittance*** or ***angle admittance relay***. It is called a mho relay because its characteristic is a straight line when plotted on an admittance diagram (*G–B* axes) as shown in Fig 5.9 (b).

A mho type relay combines both the directional and measuring functions in one unit. It is, most sensitive of all distance relays. It is least sensitive to power swings, and hence it is most suitable for phase fault relaying for long lines.

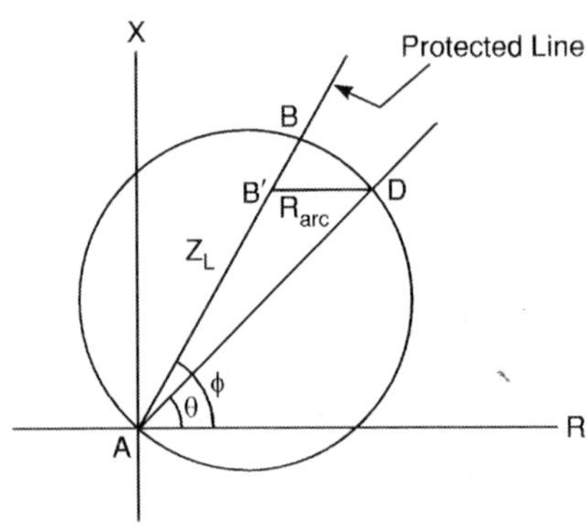

Fig. 5.10 Reduction of relay reach by fault (arc) resistance

The reach point setting of a mho relay changes with the fault angle. The fault angle depends on relative values of *R* and *X*. In case of an arcing fault condition, the value of R increases by the amount of resistance of the arc R_{arc} causing the fault angle to change. Thus, under reach is caused for settings of characteristic angle equal to the line angle. To avoid any underreach, the relay characteristics angle is set leading the line angle considering appropriate values of arc resistances. This is shown in Fig. 5.10.

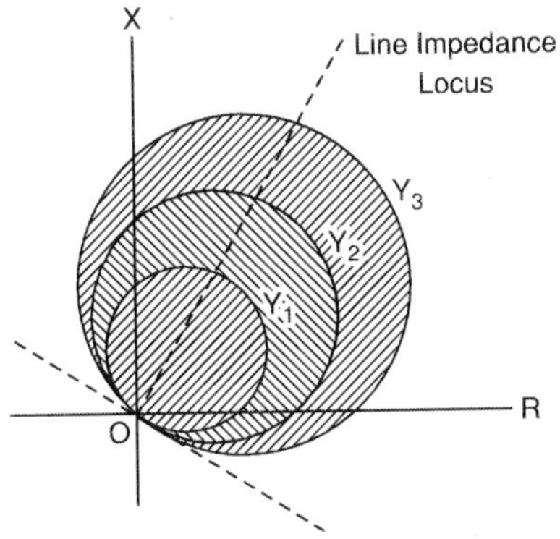

Fig. 5.11 Operating characteristics of a three-zone mho relay

Figure 5.11 shows the application of a mho relay for zone protection. It can be seen from Fig. 5.11 that a mho relay is inherently a directional relay and does not need additional directional unit. It is to be noted that if the location of an arc due to fault is just short of the point on the line at which the protective zone changes, it will cause the relay in the next zone to operate.

5.8 MODIFIED IMPEDANCE-TYPE RELAY

The modified impedance-type distance relay is like the impedance type except that the impedance-unit characteristics are shifted as shown in Fig. 5.12.

If an additional voltage proportional to the current is introduced into the voltage supply, the characteristics of the relay are modified. This additional voltage is called the ***current bias***.

The modified torque equation is

$$\tau = K_1 I^2 - K_2(V + CI)^2$$

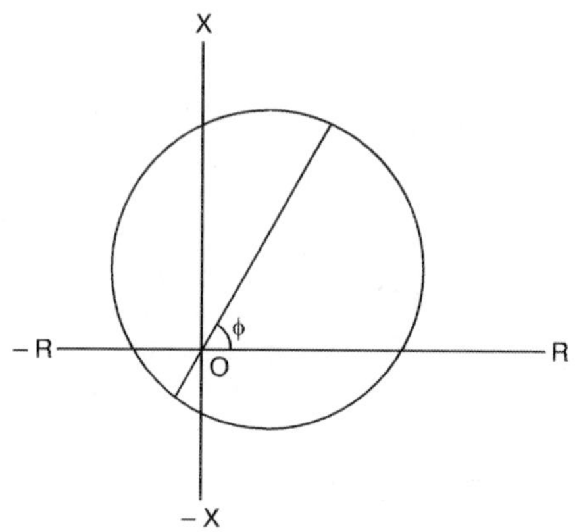

Fig. 5.12 Operating characteristics of a modified impedance-type distance relay

The term $(V + CI)$ is the rms magnitude of phasor addition of V and CI, where C is a constant.

The characteristic when plotted on the R-X diagram is a circle with centre shifted from the origin, as shown in Fig. 5.12. By such biasing, a characteristic circle can be shifted in any direction from the origin, and by any desired amount, even to the extent that the origin is outside the circle.

The offset mho (modified impedance) unit can be used in conjunction with mho measuring units as a fault detector. For mho offset unit by

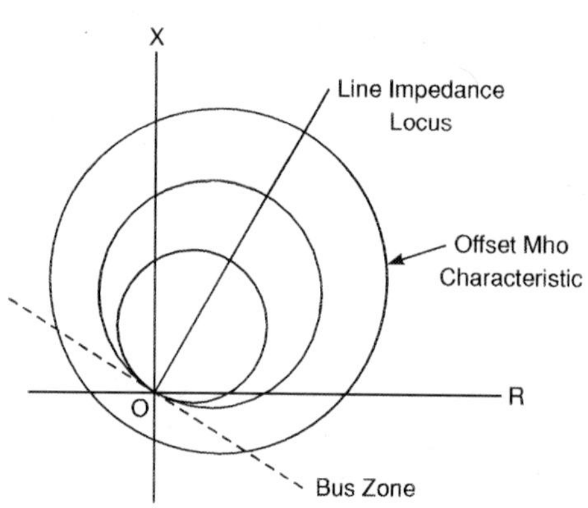

Fig 5.13 Offset mho characteristic for zone backup protection.

setting the backward reach into the bus zone provides back up protection for bus faults. This type of characteristics is shown in Fig 5.13.

5.9 OHM RELAY

An ohm relay measures a component of the impedance of the line at the relay location. The ohm relay characteristic is represented by

$$Z \cos(\phi - \theta) = \frac{K_2}{K_1} \qquad ...(5.9.1)$$

It is a straight line when plotted on the R-X plane. If $\phi \neq 90°$ in Eq (5.9.1), then the straight line characteristic is inclined to the R axis at some angle. Therefore, a reactance relay is a special ohm relay.

Ohm relays are used in conjunction with other relays as ***blinders*** to prevent distance relays from tripping on severe power swings. The ohm relay limits the area of operation on the R-X diagram of the other relay in conjunction with it. Such a use is shown in fig 5.14.

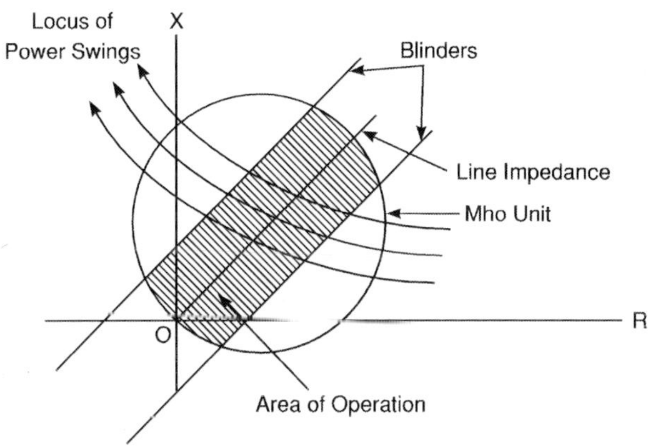

Fig. 5.14 Use of ohm relay units as blinders

The scheme usually has two ohm units connected in series with a mho unit. The characteristics are arranged to be parallel to and on the either side of the line impedance vector. As the impedance changes during a power swing, the point representing the impedance moves along the power swing locus. The ohm units operate as the power swing locus enters into the zone between the two binders. No condition other than power swing causes the impedance phasor to move in this way to operate the relay even in case of faults.

Ohm relay is also known as ***angle impedance relay***.

EXERCISES

1. Define the terms reach, underreach and overreach. Explain their significance in distance protection.
2. Explain stepped time-distance characteristics of three distance relaying units used for first, second and third zones of protection.
3. What are different types of distance relays ? Compare their merits and demerits.
4. What is an impedance relay ? Draw and explain the impedance relay characteristic on the R-X diagram.
5. Draw and explain the operating characteristic of a reactance relay.
6. Draw and explain the characteristic of a MHO relay. Why a MHO characteristics is preferred for protection of long lines against phase faults.
7. What is an angle impedance relay?
8. What are blinders ? In what circumstances, they are used in conjunction with an OHM relay ?

❑❑❑

6

TRANSFORMER PROTECTION

6.1 DIFFERENTIAL PROTECTION

A differential relay can be defined as the relay that operates when the phasor difference of two or more similar electrical quantities exceeds a predetermined value.

The differential relaying is the most selective relaying principle that detects faults by comparing the electrical quantities at all of the terminals of a system element.

Most differential relays are of the ***current differential*** type. Figure 6.1 shows a simple example. The system element to be protected is shown by the differential relay is shown by the dotted line. The element may be the winding of a generator, a transformer, a bus or a portion of a line. The CTs are placed at the ends of the element to be protected. The secondaries of the CTs are connected together and the coil of the relay is connected across the CT secondary circuit.

Consider that the current flows to a load or a fault located at point X outside the protected zone as shown in Fig 6.1 b. When the fault is outside the zone of protection, it is known as ***external fault*** or ***through fault.*** If the two current transformers have the same ratio, and are properly connected, their secondary currents will flow between the two CTs as shown by the arrows and no current will flow through the differential relay.

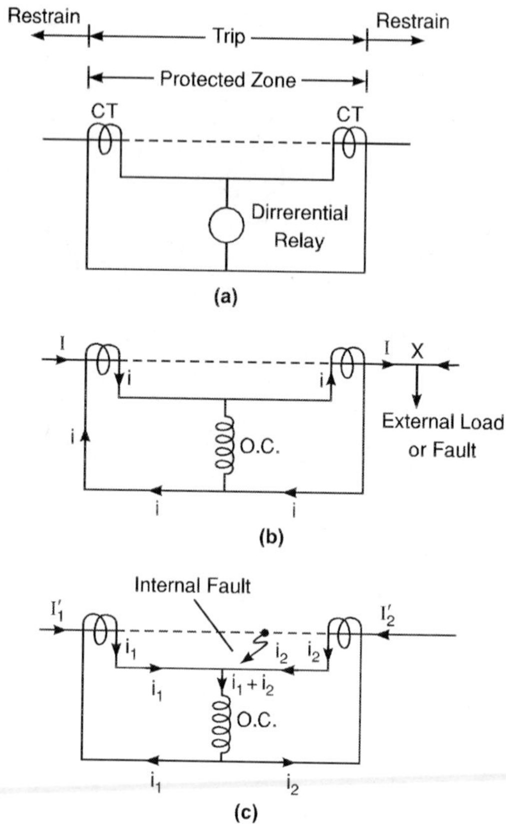

Fig. 6.1 (a) Simple differential protection, (b) Differential protection for an external fault, (c) Differential protection for an internal fault

Suppose that a short circuit develops in the protected zone of the circuit as shown in Fig. 6.1 c. For this internal fault, if the current flows to the short-circuit from both sides, as shown in Fig 6.1 c, the sum of secondary currents will flow through the operating coil O.C. of the relay, and it will operate. This form of protection is known as ***Merz-Price protection***. It is not necessary that short-circuit current flow to the fault from both sides to cause secondary current to flow the operating coil of differential relay. A flow of current on one side only, or even some current flowing out of one side, while a larger current enters the other side will cause a differential current, In other words, the differential-relay current will be proportional to the phasor difference between the currents entering and leaving the protected circuit. If the differential current exceeds the relay's pickup value, the relay will operate.

The principle can be extended to a system element having several connections. Figure 6.2 shows a three terminal current-differential application. It is necessary that all the CTs have the same ratio and they should be connected so that there is no current in the differential relay when the phasor current entering the circuit is equal to the phasor current leaving the circuit.

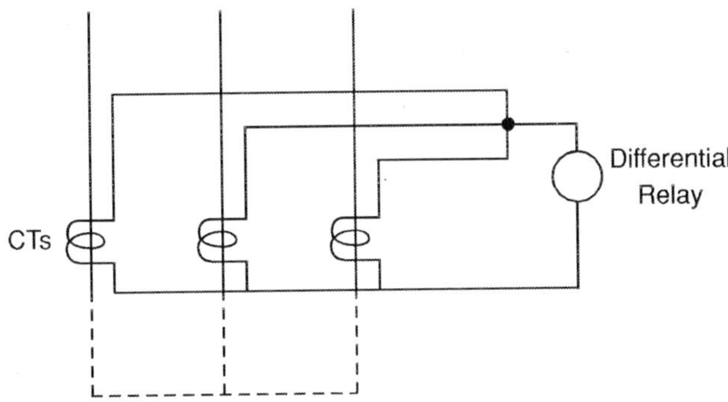

Fig 6.2 A three-element Current-differential application

6.2 PERCENTAGE DIFFERENTIAL PROTECTION

The two current transformers used with the simple differential protection are not ideal. No two CTs will function exactly alike even though they are made of the same specifications and material. Therefore they may saturate unequally when large currents flow through them during external faults. Then the two CTs will transform their primary current differently. Thus, a current flows in the operating coil due to the CT errors during an external fault.

In order to overcome these difficulties, Merz-Price protection is modified by biasing the relay. This is commonly known as ***biased differential protection*** or ***percentage differential protection***. It is shown in Fig 6.3a.

The percentage differential relay has an operating coil and a restraining coil. The operating coil is connected to the midpoint of the restraining coil. The purpose of the restraining coil is to prevent undesired relay operation in case a current flows in the operating coil due to CT errors during an external fault.

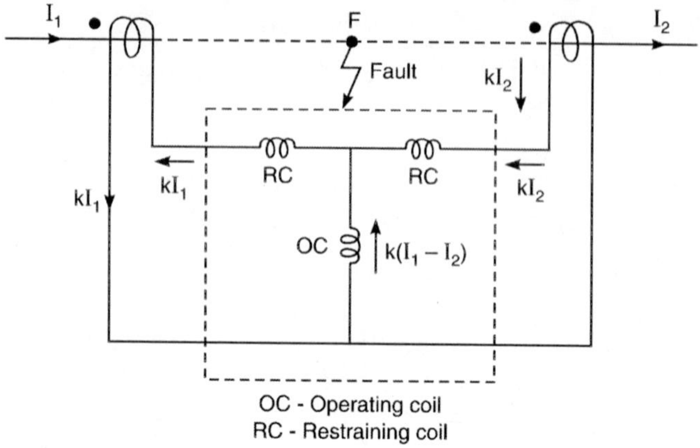

Fig. 6.3a Percentage differential protection

The differential current in the operating coil is proportional to $(I_1 - I_2)$ and the equivalent current in the restraining coil is proportional to $\frac{1}{2}(I_1 + I_2)$ due to the fact that the two parts of the restraining coils are identical

Let

N_0 = number of turns of the operating coil

N_r = number of turns of the restraining coil

τ_0 = torque developed by the operating coil

τ_r = torque due to restraining coil

Since the torque developed is proportional to the ampere turns

$$\tau_0 \propto |(I_1 - I_2)|\, N_0 = k\,|(I_1 - I_2)|\, N_0$$

$$\tau_r \propto |(I_1 + I_2)|\, N_r = k\,\left|\frac{1}{2}(I_1 + I_2)\right|\, N_r$$

The relay will operate when $\tau_0 > \tau_r$, that is,

$$k\,|(I_1 - I_2)|\, N_0 > k\,\left|\frac{1}{2}(I_1 + I_2)\right|\, N_r$$

or

$$\frac{|I_1 - I_2|}{\left|\frac{1}{2}(I_1 + I_2)\right|} > \frac{N_r}{N_0}$$

This relation shows that the relay will operate when the magnitude of the phasor difference in the currents at the two ends of the protected element exceeds a certain fixed percentage $\left[\left(\frac{N_r}{N_0}\right) / 100\right]$ of half the magnitude of their phasor sum. That is why this relay is called ***percentage differential relay***.

The relay actually operates when

$$k\,|(I_1 - I_2)| - \left(\frac{N_r}{N_0}\right) k \left|\frac{I_1 + I_2}{2}\right| > |I_p|$$

where $|I_p|$ is the predetermined value of current.

Since the difference in currents in a percentage differential relay has to be much larger than $|I_p|$, under healthy conditions of the protected element the CT errors cannot cause undesired tripping, as always

$$|k_1 I_1 - k_2 I_2|_{\text{Through fault}} < \left(\frac{N_r}{N_0}\right) k \left|\frac{I_1 + I_2}{2}\right|$$

where k_1 and k_2 are the transformation ratios of the two CTs due to errors. Also, $k_1 = k_2 = k$ where no CT error is considered.

The operating characteristic of a percentage differential relay is a straight line with a slope of $\left(\frac{N_r}{N_0}\right)$ and an intercept equal to minimum pickup current I_{pu} on the y axis. All points above the straight line represent the condition where the operating torque is greater than the restraining toque and hence fall in the **trip region** of the relay. All points below the straight line belong to the **restraining region**. The operating characteristic of the percentage differential relay is shown in Fig. 6.3b.

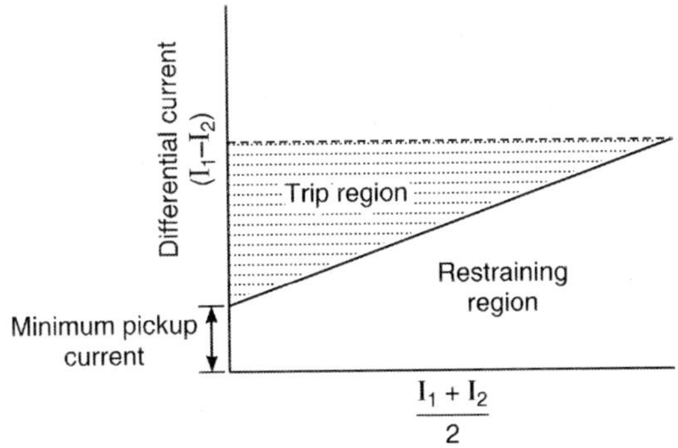

Fig. 6.3b Operating characteristic of a percentage differential relay

The slope of the relay is usually expressed as a percent value : typically 10, 20 and 40%. A relay with a slope of 10% is more sensitive than a relay with a slope of 40%.

It is to be noted that if the fault current does not flow out of the protected zone, the current I_1 entering at one end of the element is equal to the current I_2 coming out at the other end. In this case the differential relay protection is ineffective. An example of this type of fault is an interturn fault on the generator phase winding. Other methods are to be used for protecting of such types of faults.

6.3 TRANSFORMER PROTECTION

The protective scheme chosen for power transformer circuits depend on the size and importance of the unit and its relation to the system as a whole. For small units ordinary overcurrent protection schemes may be employed and if desired earth fault protection may be added. For large units (1 MVA and above) percentage differential schemes are universally adopted. Sometimes small transformers are included with transmission line into a single protective zone. Provisions for magnetizing inrush currents, phase shifts and transformer ratio are taken care of in differential schemes.

Transformer protection can be divided into two main classes.

(a) Protection of the transformer against the effects of faults occurring on any part of the system. These are known as *External faults*.

(b) Protection of the system against the effects of faults arising in a transformer. These are known as *Internal faults*.

External Faults

Abnormal conditions due to external factors result in stress or damage to the transformer. These conditions may be:

1. ***Overloading*** which results in increased copper losses and temperature rise.
2. ***Overvoltages,*** either power frequency or transient voltage surges produce increased insulation stress and increased flux leading to more core losses and possible saturation of the core.
3. ***System faults*** generally produce large currents in the system, hence overloading the transformer.
4. ***Under frequency*** operation of the system causes increased flux which results in increased core losses and temperature rise.

Internal Faults

The fault conditions which may arise with in a power transformer include

1. Phase to phase faults
2. Earth faults
3. Inter-turn faults

These faults cause flow of the currents that depend on the source, neutral grounding impedance, winding connections, leakage reactance of transformer, and the position of the fault etc. These faults can cause immediate and permanent damage to the transformer. Such faults can be detected at the winding terminals by imbalances in the voltages and currents in the windings.

In addition to above faults *core faults* may occur due to breakdowns of insulation between the core laminations giving rise to sufficient eddy current flow to cause overheating of the core and windings. Such faults are not detectable from the usual current and voltage imbalances at the winding terminals. Core faults are generally the result of *incipient faults* which are initially minor in nature but slowly develop into major faults. They may be due to deterioration of insulation due to ageing or sustained overheating.

Faults on auxiliary equipments which is a part of the transformer can be considered as internal faults. They include failure of cooling air or oil pumps, tank oil leakage or deterioration of insulating properties of transformer oil etc.

In general, a transformer may be protected by fuses, overcurrent relays, differential relays and pressure relays and temperature detecting devices. The choice of protection depends on several factors such as MVA rating, location, voltage, connection and design.

6.4 PROTECTION AGAINST MAGNETIZING INRUSH CURRENT

When an unloaded transformer is switched on, it draws a large magnetizing current to produce the core flux. This initial magnetizing current is called *magnetizing inrush current*. This current can have a peak of 8 to 10 times the normal full load peak current of the transformer. This current has no equivalent on the secondary side of the transformer. Maximum inrush occurs if the transformer is switched in when the supply voltage is zero, that is, the inrush current flows only in the primary winding.

Magnetizing inrush current is characterized by large harmonic components which are not noticeably present in the short-circuit current. That is, for faults, the harmonics are of low values. The ***second harmonic*** component is more in the inrush current than in the fault current. This feature can be utilized to distinguish between the inrush currents and fault currents.

As the inrush current flows only in the primary winding. This current will appear as an internal fault in the differential protection. This inrush current will be the out-of-balance current and can trip the transformer.

Therefore, it is important for the design of the differential relays for transformer protection, that no tripping takes place due to the magnetizing inrush currents.

A number of schemes based on the harmonic properties of the magnetizing inrush current are used to prevent tripping due to large magnetizing inrush currents. A relay circuitry is used which is designed to prevent undesired tripping due to magnetizing inrush currents.

Two such methods used in practice are as follows:

(a) Providing a protection scheme with harmonic current restrain.

(b) Temporary desensitizing or disconnecting differential relays during the period of inrush of magnetizing current.

6.5 METHOD OF HARMONIC CURRENT RESTRAINT

This method of protection utilizes the harmonics of the magnetizing inrush current to prevent operation. Magnetizing inrush current is characterized by large harmonic components which are not noticeably present in the short-circuit current. That is, for faults the harmonics are of low value. This feature is utilized to distinguish between a fault current and magnetizing inrush current. The ***Second Harmonic***, which is a major component in the inrush wave, is used to restrain or prevent relay operation. In some designs it is used alone, while in others in combination with the harmonics. The harmonic components in the inrush current are filtered out and introduced into a circuit that restrains the relay from operating. Such a relay is called *harmonic-current-restraint percentage differential relay.*

Figure 6.4 shows the harmonic-current-restraint differential relay protection of a star-delta connected power transformer.

The restraining coil of the relay is fed with the sum of the fundamental and harmonic current components from the *through-current* transformer. The operating coil of the relay receives from the differential-current transformer only the fundamental component of the

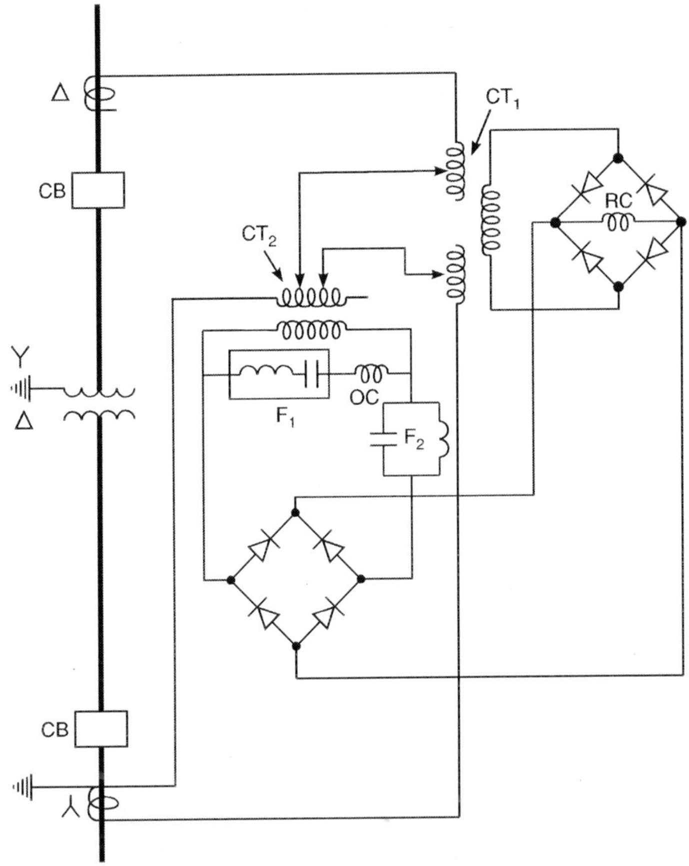

CT$_1$ – Through-Current Transformer
CT$_2$ – Differential Current Transformer
F$_1$ –Filter (Passes only Fundamental)
F$_2$ – Filter (Blocks only Fundamental)
OC – Operating coil
RC – Restraining coil

Fig. 6.4 Harmonic-current-restraint percentage differential relay

differential current, the harmonics being separated, rectified, and fed back in the restraining coil. The direct-current component, present in both magnetizing inrush current and offset fault current is largely blocked by the differential-current transformer and the through-current transformer, and produces only a slight momentary restraining effect.

Under normal conditions during energization, the restrain torque will always be greater than the operating torque. That is, during charging, the operating coil only gets the fundamental F and the restraining coil gets $(F + H)$ from the through-current transformer and another rectified harmonic component h. The relay does not pick up because

$$(F + H + h) >> F$$

In case of a fault or short circuit, the harmonics are of low value. The fundamental component of current flows through both operating coil and restraining coil and the relay is set to operate for that specified value of fundamental component.

6.6 CONSIDERATIONS TO PROTECT A THREE-PHASE TRANSFORMER BANK

Differential protection is the main scheme use for transformers. A number of considerations should be made to protect a three-phase transformer bank

1. We must take into account the change in current magnitude and phase in going from low voltage side to high voltage side. The differences must be cancelled out before being applied to the differential relay.
2. The current transformers should have ratings to match the rated currents of the transformer winding to which they are applied.
3. In case of a star-delta transformer, there is a phase shift of 30° in line currents. The sequence current flowing in the star side of the transformer does not produce current outside the delta on the other side. Therefore, the zero sequence current should be eliminated from the star side. This condition is also fulfilled by the C.T. connected in delta in the star side of the transformer. *The current transformers should be connected in star for a delta winding and in delta for a star winding to provide the* 30° *phase shift.* If the power transformer were connected in star-star, the CTs on both the sides should be connected in delta. When the CTs are connected in delta, their secondary ratings should be reduced to $(1/\sqrt{3})$ times the secondary ratings of the star connected CTs in order that the currents outside the delta may balance with the secondary currents of the star-connected CTs.
4. Allowance should be made for tap changing by providing restraining coils (bias). The bias should exceed the effect of maximum deviation ratio.

6.7 DESENSITIZING SCHEME FOR PROTECTION OF TRANSFORMERS

This scheme of protection makes the differential relays temporarily insensitive (but only at full voltage) during the inrush period of magnetizing current. Fig. 6.5 shows this method. It uses two voltage relays namely, time delay voltage relay (TDVR) and high-speed voltage relay (HSVR). These relays are supplied from a voltage transformer (VT).

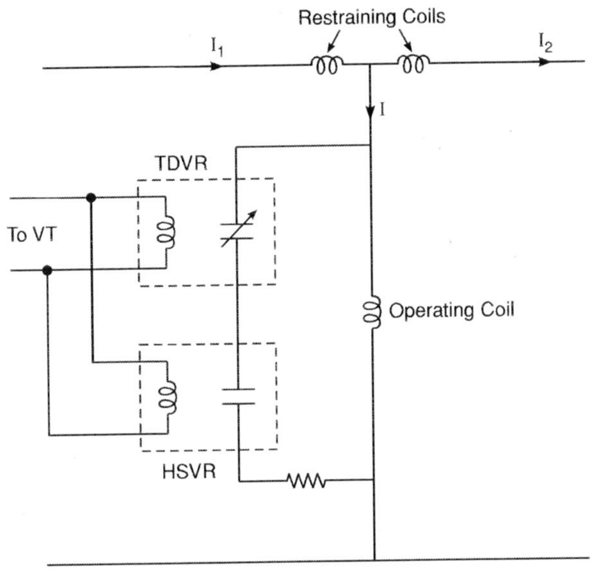

Fig. 6.5

The two voltage relays control the shunt path to the operating coil. The time delay voltage relay has normally closed contact and the high-speed relay are normally open. When the power transformer is switched on (no fault), the contacts of high speed voltage relay are closed and the contacts of time delay voltage relay do not open. Thus, the shunt path of the operating coil of the percentage differential relay is closed and the operating coil does not operate. After a short time, the magnetizing current decays to steady magnetizing current and the high speed voltage relay contacts open. Consequently, the percentage differential relay which was temporarily desensitized during the magnetizing inrush current period becomes operative again against internal faults. If the power transformer is switched on to a fault, the voltage is low that the high speed voltage relay does not pick up, and the shunt path remains open and the percentage of differential relay remains operative.

6.8 BUCHHOLZ RELAY

Buchholz relay is a gas pressure actuated relay for alarm and tripping. It is used in addition to differential relay and over current protection. It

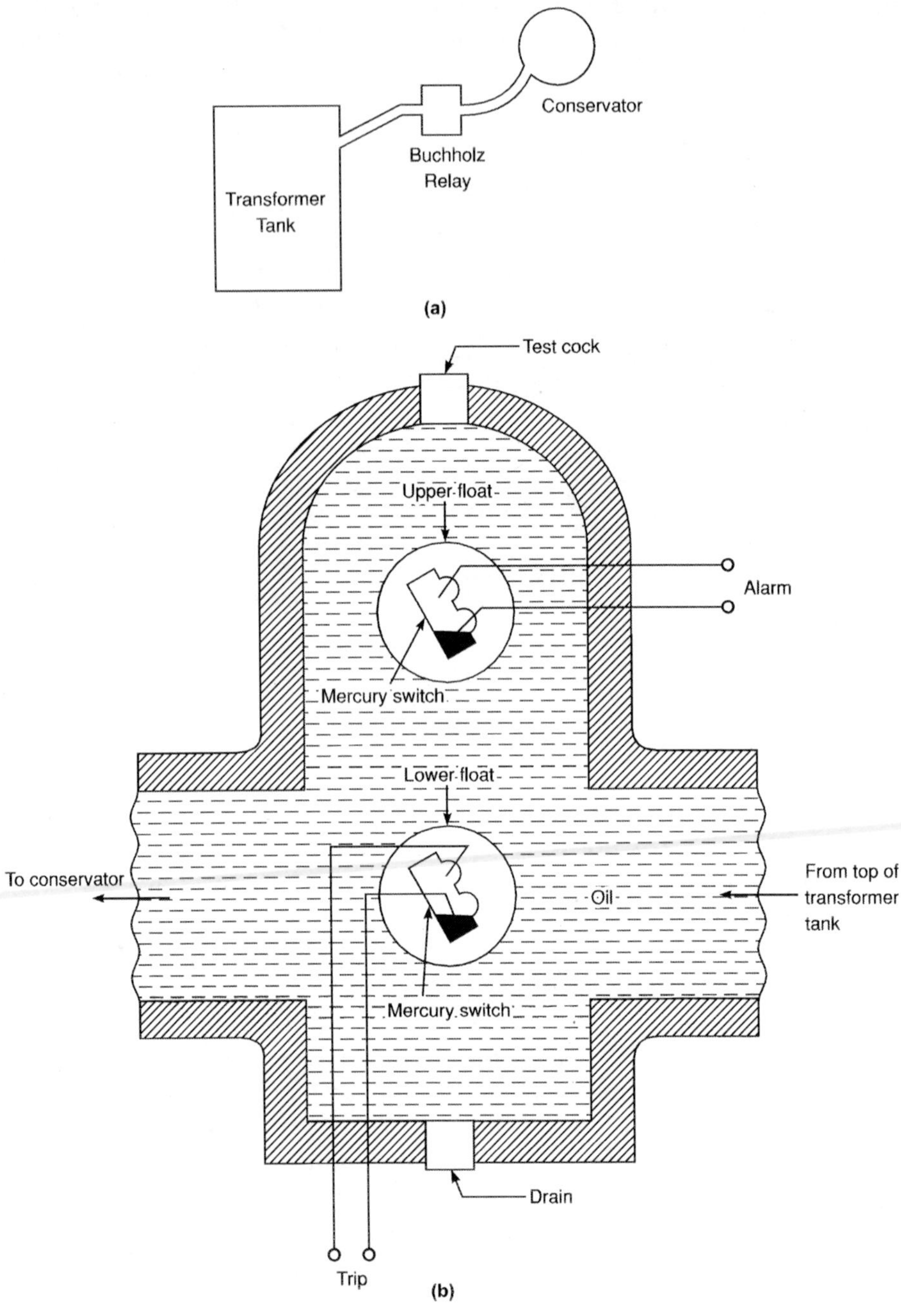

Fig 6.6 Buchholz relay

is the simplest form of relay used only for transformers provided with conservators. The relay is connected between top of the transformer tank and the conservator by means of pipe as shown in Fig 6.6. a. It consists of two hinged hollow floats in a metallic chamber. The transformer tank is filled with oil. One float is near the top of the chamber and the other is in the path of the oil between the tank and conservator. Inside these floats, mercury switches are provided as shown in Fig. 6.6 b.

When a fault develops within a transformer, the oil of the tank is overheated decomposing the oil. The decomposition produces gases such a hydrogen, carbon monoxide and light hydrocarbons. In case of minor faults the gas bubbles rise to the top of the chamber and the oil level falls down. The upper float is lowered and the alarm circuit is closed through the mercury switch to sound an alarm. The sample of the accumulated gas can be collected through the test cock. Its chemical analysis is performed to know the type of fault.

When a major fault occurs, large volumes of gases are generated and the violent displacement of oil takes place. The rush of oil operates the valve to close contacts that trip the transformers circuit breakers.

Buchholz relay provides an alarm for a number of fault conditions including:

1. Inter turn faults or winding faults involving only lower power levels.
2. Core hot spots due to short circuits on the lamination insulation.
3. Faulty joints.
4. Core bolt insulation failure.

Major faults resulting in tripping include severe earth or inner phase winding faults and the loss of oil through leakage.

6.9 OVERFLUXING PROTECTION

The emf equation of a transformer is given by

$$E = 4.44\, \phi_m f T$$

$$\phi_m = \frac{E}{4.44 f T}$$

Thus, at constant frequency, whenever there is an overvoltage, the magnetic flux increases. Transformers are normally subjected to flux levels approaching the knee point in their magnetizing characteristic. Typically, the rated voltage at rated frequency may be 10% below the saturation level. If the core flux exceeds the situation level, the transformer draws an excessive magnetization current. This condition is called **overexcita-**

tion. Under this condition, the flux patterns in the core and surrounding structures change. Significant flux levels may be reached in the transformer tank and other structural members. As these are not laminated, very high eddy currents are likely to result, producing severe damage to the transformer. It can be seen that overexcitation can also occur in case of low-frequency operation of the transformer at rated voltage. Therefore, to keep the working flux within the permissible design limits, the (E/f) ratio must not exceed the permissible limit. The overexcitation can be detected by measuring the (E/f) ratio by a so-called **volts per hertz relay**. This protection is called **volts/hertz protection**.

This protection is specially needed in case of unit-connected generator transformers. The transformer should be tripped in one or two minutes at the most if overfluxing persists.

Microprocessor-based volts/hertz relays have been developed and are in use. Many volts/hertz relays have two settings, a lower setting for alarm and a higher setting for tripping.

EXERCISES

1. Explain clearly the basic principle of operation of a percentage differential relay for a) internal fault, b) external fault.
2. Explain with reasons the connection of CTs for protection against internal faults in a delta-star transformer.
3. What is magnetizing inrush current ? How it is distinguished from fault currents? Suggest a protective scheme which protects a transformer against faults but does not operate in case of magnetizing inrush current.
4. Explain the function of Buchholz relay in the protection of transformer.
5. What protective devices other than differential protection are used for protection of a large transformer? Describe briefly.
6. What are the different types of faults occurring on a transformer? Give their causes, effects and remedies.
7. Give the considerations made in protection of a three-phase transformer banks.
8. Explain with a neat diagram the method of harmonic-current restraint for the protection of a transformer.
9. Explain desensitizing scheme for protection of transformers.

❑❑❑

7

GENERATOR AND MOTOR PROTECTION

7.1 INTRODUCTION

The following are the various types of faults that can occur in an ac generator (alternator):

Stator Faults

(a) Phase-to-phase faults
(b) Phase-to-ground faults
(c) Inter-turn faults

Rotor Faults

(a) Rotor earth fault
(b) Loss of excitation
(c) Rotor overheating because of unbalanced three-phase stator currents

Miscellaneous Abnormal Conditions

(a) Overvoltage
(b) Over heating
(c) Over speed
(d) Failure of prime mover

7.2 PERCENTAGE DIFFERENTIAL PROTECTION

Figure 7.1 shows the protection of a three phase star-connected generator by means of a percentage differential relay. Current transformers (CTs) in star are provided on both the outgoing side and the generator winding connection to ground. The restraining coils are energized by the secondary connections of current transformers in each phase. The operating coils are connected from the tapings on the restraining coils and the ground connection.

In case of a fault due to a short circuit in the protected zone of the generator windings, the currents in the primary windings of current transformers on both sides of the generator winding of the same phase become different. This results in a difference between the secondary currents of the two CTs. These currents flow through the operating coils and actuate the relays to trip the main circuit breaker.

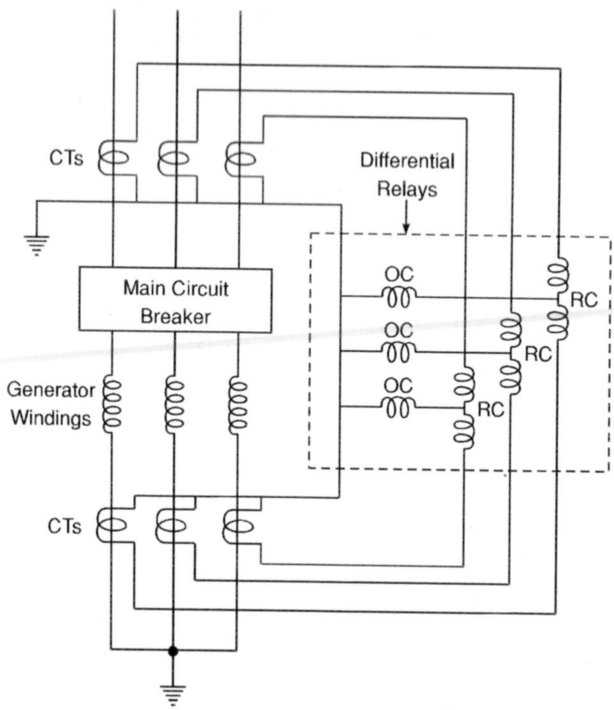

Fig 7.1 Percentage differential protection of a star-connected generator

For stator faults the percentage differential relay also operates an auxiliary relay which initiates the following operations:

(a) It trips the main circuit breaker of the generator.
(b) It trips the field circuit breaker.
(c) It trips the neutral circuit breaker, if there is one.
(d) It shuts down the prime mover.
(e) It turns on CO_2 gas if provided for safety of the generator.
(f) It operates an alarm.

The percentage differential scheme does not provide protection against external faults or overloads. It provides complete protection against phase to phase faults. It provides protection against ground faults to about 80 to 85 percent of the generator windings. It does not provide protection to 100 percent of the winding because it is influenced by the magnitude of the earth fault current which depends upon the method of neutral grounding.

Example 7.1. A 3-phase, 20 MVA, star-connected generator is protected by the current balance system of protection. If the ratio of CTs is 1200/5, the minimum operating current of the relay is 0.75 A and the neutral resistance is 6Ω, calculate the percentage of each phase of stator winding which is unprotected against earth faults when the machine is operating at normal voltage.

Solution : Minimum fault current to operate the relay

$$= \text{CT ratio} \times 0.75$$

$$= \frac{1200}{5} \times 0.75 = 180 \text{ A}$$

Let a% of the winding is left unprotected

The voltage induced in the unprotected section of the winding

$$= a\% \text{ of the phase voltage}$$

$$= \frac{a}{100} \times \frac{11000}{\sqrt{3}} = 63.5a$$

$$\text{Fault current} = \frac{\text{voltage induced}}{\text{earth resistance}} = \frac{63.5a}{6}$$

For the relay to operate

$$\frac{63.5a}{6} = 180$$

$$a = \frac{6 \times 180}{63.5} = 17$$

Example 7.2. A 13.8 kV, 125 MVA, star-connected alternator has a synchronous reactance of 1.4 per unit per phase and negligible resistance. It is protected by a Merz-Price balanced current system which operates when out of balance current exceeds 10% of the full-load current. If the neutral point is earthed through a resistance of 2Ω, determine what portion of the winding is protected against earth fault.

Solution : Let a% of the winding be unprotected.

$$\text{MVA} = \sqrt{3}\,\text{V}_\text{L}\text{I}_\text{L} \times 10^{-6}$$

$$125 = \sqrt{3}(13.8 \times 10^3)I_L \times 10^{-6}$$

$$I_L = \frac{125 \times 10^6}{\sqrt{3} \times 13.8\ 10^3} = 52229.6 \text{ A}$$

$$\frac{I_p X_\Omega}{V_p} = X_{pu}$$

where,

I_p = phase current

V_p = phase voltage

X_Ω = reactance per phase in ohms = 1.4 Ω

X_{pu} = per unit reactance

In star connection, phase current = line current

$$\therefore \quad I_p = I_L$$

$$V_p = \frac{V_L}{\sqrt{3}} = \frac{13.8 \times 10^3}{\sqrt{3}}\ V$$

$$\frac{5229.6\ X_\Omega}{\left(\dfrac{13.8 \times 10^3}{\sqrt{3}}\right)} = 1.4$$

∴ Reactance per phase

$$X_\Omega = \frac{1.4 \times 13.8 \times 10^3}{\sqrt{3} \times 5229.6}\ 2.133\ \Omega$$

Voltage induced in unprotected portion of the winding

$$= a\%\text{ of } V_p$$

$$\frac{a}{100} \times \frac{13.8 \times 10^3}{\sqrt{3}} = 79.67\ a$$

Earth fault current = 10% of full load current

$$\frac{10}{100} \times 5229.6 = 522.96 \text{ A}$$

Reactance of the unprotected winding $= a\%$ of X_Ω

$$\frac{a}{100} \times 2.133 = 0.02133a$$

Earthing resistance = 2Ω

$$\text{Earthing impedance} = \frac{\text{voltage induced}}{\text{earth fault current}}$$

$$\frac{79.67a}{522.9} = 0.1523a$$

Since

$$(\text{earthing impedence})^2 = (\text{earthing resistance})^2 + (\text{earthing reactance})^2$$

$$(0.1523a)^2 = (2)^2 + (0.02133a)^2$$

$$a = 13.26\%$$

Therefore protected winding = 100 − 13.26 = 86.74%

7.3 STATOR EARTH FAULT PROTECTION

To protect against stator earth faults, the neutral point of the generator is connected to earth through a high resistance. A voltage transformer (VT) is connected across the resistance between the neutral point and ground. The voltage transformer is connected to a polarized voltage relay. In case of an earth fault, current will flow through the resistance, producing a potential difference across the resistance. This voltage difference is applied to the voltage transformer connected to the relay.

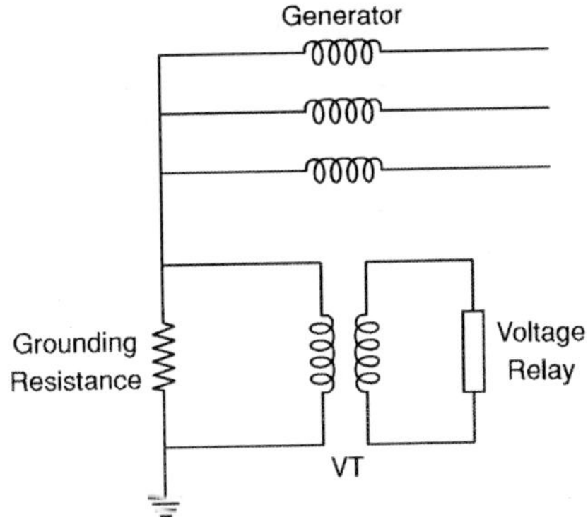

Fig. 7.2 Stator earth fault protection

The output of the voltage transformer actuates the relay which trips the generator circuit breaker as shown in Fig 7.2.

This protection system is designed for a fault current exceeding 10A.

7.4 STATOR INTERTURN FAULT PROTECTION

Current leakage can take place between turns of the same phase of a winding or between parallel coils of the same phase. This is known as an *interturn fault.*

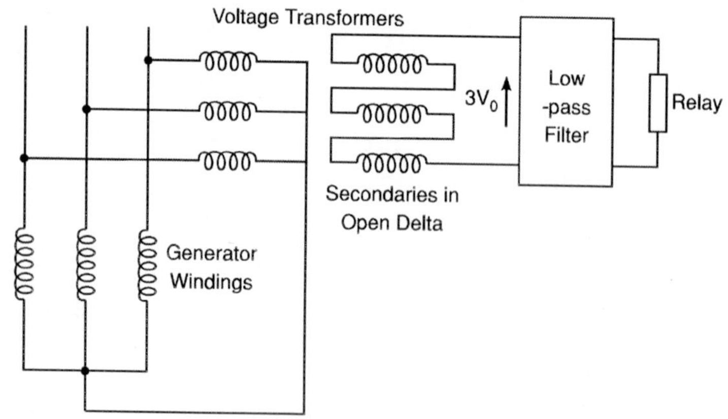

Fig 7.3 Interturn fault protection scheme

Figure 7.3 shows an interturn fault protection scheme. This is based on zero-sequence voltage measurement across the generator. In generator with one winding per phase, a voltage transformer (VT) is connected between each phase terminal and the neutral of the winding. The voltage transformer is used to extract zero sequence component. The secondary winding of the voltage transformer is connected in open delta to provide the zero-sequence voltage to a polarized voltage relay.

The zero-sequence voltage does not exist during normal operating conditions. In the event of an interturn fault, the generated emf contains zero-sequence component. A voltage appears at the terminals of the open-delta. A filter is provided to extract a third harmonic component from the output of the voltage transformer. The zero-sequence voltage is applied to the relay and causes it to trip.

If V_0 is the zero-sequence voltage

$$V_1 + V_2 + V_3 = 3V_0$$

The voltage across the terminals of the open delta is $3V_0$.

Large turbo alternators usually have one turn per phase per slot. Therefore interturn fault protection is not required in such alternators. Interturn fault protection is provided in hydrogenerators which have parallel windings in each phase. Each phase is divided into two equal groups of parallel circuits with a CT for each group. For generators having odd number of parallel circuits per phase, suitable CT ratios provide the required current balance. This is known as *Split-phase relaying*. Figure 7.4 shows split phase relaying for a multi circuit generator.

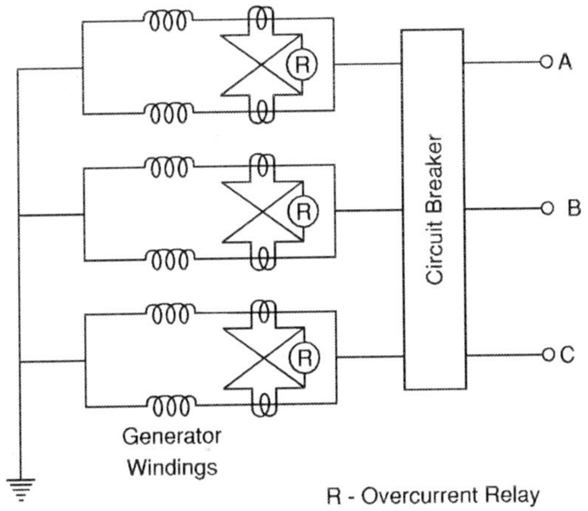

Fig 7.4 Split-phase relaying for a multi-circuit generator

7.5 ROTOR EARTH FAULT PROTECTION

The field circuit of an alternator is operated ungrounded therefore in case of a single ground fault there is no flow of current from rotor to ground. Consequently, the operation of the generator is not affected or causes any damage to the generator. If a ground fault has occurred on the field winding it increases the stress to ground at other points in the field winding due to voltages induced in the field winding by stator winding transients. Thus, increasing the probability of a second ground fault may bypass a part of the field winding, increasing the current through the remaining portion of the winding. This will cause an imbalance in the air gap fluxes and thus an imbalance of magnetic forces on opposite sides of the rotor. These forces are large enough to make the motor vibrate and make it eccentric leading to severe mechanical damages.

In some cases a second ground fault may not bypass enough portion of the field winding to cause such a large imbalance in magnetic forces. Arcing due to faults may cause local rotor heating slowly leading to temperature rise and associated problems.

Figure 7.5 shows a method of protection of rotor against earth faults or open circuits. A dc voltage is applied between the field circuit and ground.

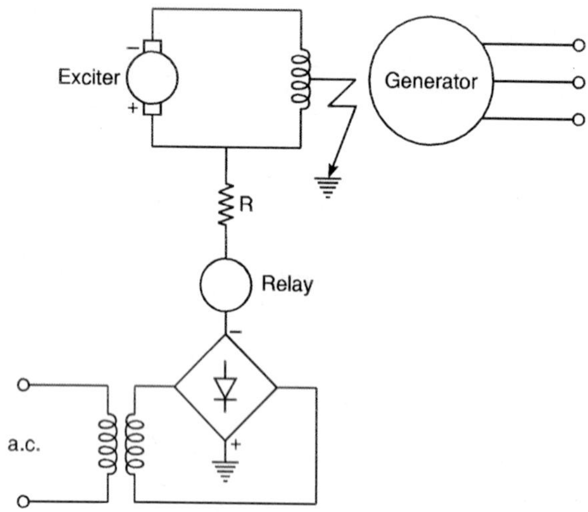

Fig 7.5 Rotor earth fault protection

A fault detecting relay (polarized moving iron type) and a high resistance to limit the connected are connected in series as shown. A fault at any point on the field circuit will make a current of significant magnitude to flow through the relay, causing its operation. For a single earth fault usually only alarms are sounded. Load is removed from the generator as soon as possible and the generator is shut down for fault rectification. The main field winding circuit is not accessible in brushless machines. Any field fault is detected in the exciter control circuit. Vibration detecting equipment must be provided for possible detection of rotor earth faults.

7.6 LOSS OF EXCITATION

A synchronous generator may either operate alone or it may be connected in parallel with other generators. If the excitation fails in an isolated generator, it stops supplying power to the load.

In case of failure of excitation of a generator connected to a grid, it operates as an induction generator running above synchronous speed.

The generator operating in parallel with other generators may either be salient pole type or it may be cylindrical rotor type. In cylindrical rotor machines overheating of the rotor takes place due to heavy rotor currents in rotor iron. The rotor of salient pole generator is not overheated because it has damper windings to carry induced currents. The stator of the alternator gets overheated due to the magnetizing current drawn by the machine from the system. The generator may be supplying reactive power to the system before the loss of excitation. After the failure of excitation the generator receives reactive power from the system. Thus, this large reactive load is suddenly thrown on the system, in addition to the loss of generator's reactive power output. This may provide large reduction in system voltage. Consequently the system may become unstable, unless the other generators can automatically pick up the additional reactive load immediately.

Undercurrent relays in the field circuit are widely used for protection against loss of excitation. Directional distance relays at the generator terminals are preferred for more selective operation. Figure 7.6 (a) shows the relay connections for loss of excitation.

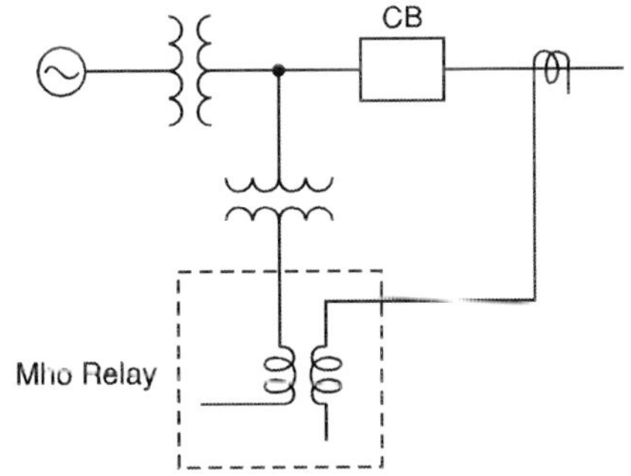

Fig 7.6 (a) Relay connections for loss of excitation

The relay setting is so adjusted that it operates whenever the excitation goes below a certain value. For large modern generators offset mho relay is used for the protection against loss of excitation. With the loss of excitation of the generator, the locus of equivalent generator impedance moves from the first quadrant to the fourth quadrant on the R-X diagram as shown in Fig 7.6 (b) irrespective of the initial conditions. This locus is not traced in any other condition.

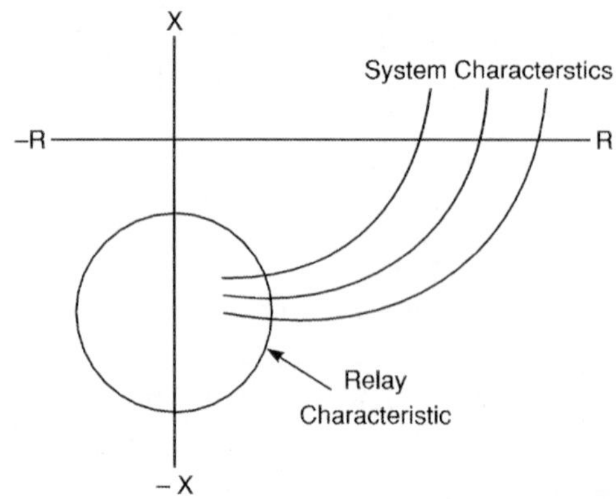

Fig 7.6 (b) Loss of excitation and relay system characteristics

The relay trips the field breaker and generator is disconnected from the system.

It is to be noted that with loss of excitation the flow of reactive power is reversed and the generator impedance changes from inductive to capacitive.

7.7 PROTECTION AGAINST UNBALANCED CURRENTS

In case of fault, there is an imbalance in the three phase stator currents or voltages. The unbalanced voltages at stator terminals can be resolved into their corresponding symmetrical components. The positive sequence voltages produce a resolving field that revolves in the same speed as the rotor. Since there is no relative motion present, there are no induced voltages or currents in the rotor due to positive sequence components. Negative sequence component produces a negative field revolving the direction opposite to the rotor at a relative speed twice the rotor speed inducing large double frequency voltages and currents in the rotor. These induced currents in the rotor winding and core cause overheating of the rotor and possible damage to the generator. In addition to this, unbalanced currents may cause vibrations on the rotor leading to mechanical damages.

Fig 7.7 shows an arrangement of protection for unbalanced stator currents. Negative sequence quantities can be detected by negative sequence filters. A relay is set to operate if the negative sequence component is in excess. The unit's main breaker may be tripped. For small unbalance only some alarms are set on. The relay used in this scheme is inverse time overcurrent relay of predetermined limits.

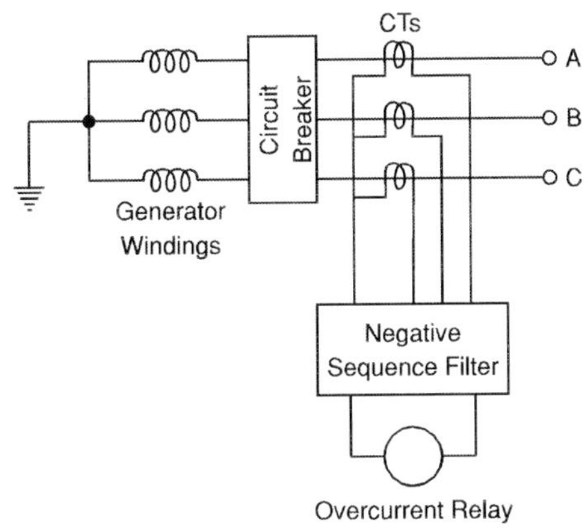

Fig 7.7 Generator protection against unbalanced currents

7.8 OVERVOLTAGE PROTECTION

Excessive stator overvoltage may occur due to sudden loss of large electrical load or a defective automatic voltage regulator (AVR). In case of a sudden loss of a large electrical load, there is an increase in the speed, thereby increasing the stator output voltage. Generally the exciter-generator-voltage regulator loop has a response fast enough to obviate the need for additional voltage control. Overvoltage protection is normally provided for hydro electric and gas turbine generators. In steam powered turbines, it is possible to bypass the steam before it leads to over speeding of the rotor. An overvoltage relay is set to operate at value of 110% of the normal value and operates instantaneously for values above 130% of the rated voltage. The operation of relay should first insert an additional resistance (variable) in the generator field circuit, and if even then the overvoltage persists, the main generator breaker and the generator or the exciter field breaker trips.

Over frequency relays set to respond to about 5% above the full load speed.

7.9 PROTECTION AGAINST OVERHEATING

Stator of an alternator may be overheated due to overloading, failure of cooling systems or core faults. An overcurrent relay may detect an overload but may fail to detect failure of cooling system or heating due to core faults. Temperature sensing elements like thermisters, thermocouples or resistance temperature indicators embedded in the stator slots are used to monitor the temperatures of the windings. A stator overheating relaying scheme is shown in Fig 7.8. Under normal conditions the bridge is balanced. Any change in temperature above the set limits unbalances the bridge and the unbalanced currents energize the relay.

Measurements of the inlet and outlet temperatures of the cooling medium are also done for detection of possible temperature rise inside the machine.

Temperature sensing elements are generally not embedded in the rotor windings because of complicated connections. Rotor temperatures rise can be estimated by measuring the rotor winding resistance.

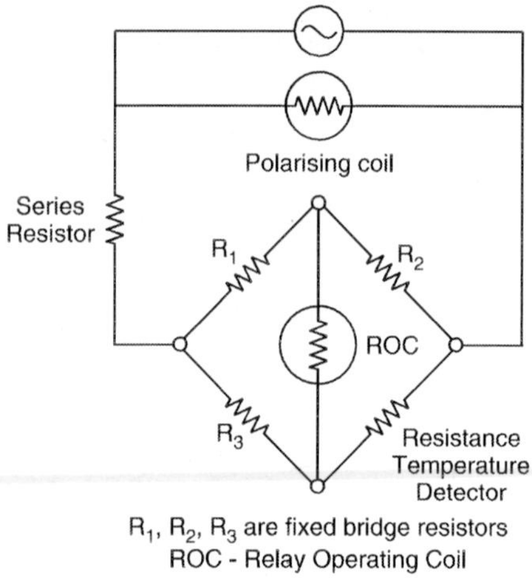

Fig. 7.8 Protection against stator overheating

7.10 PROTECTION AGAINST MOTORING (PRIME MOVER FAILURE)

Protection against motoring is beneficial to the prime movers and the system as a whole, but not to the alternator. It is usually undesirable to allow real power flow into the generators because of the hazard involved to the generator prime mover. The problems involved are

(i) Turbine blade overheating in steam turbines
(ii) Cavitation with hydraulic turbines

(iii) Backfire and fire hazard with diesel engines and gas turbines.

This problem can be corrected with a directional unit that detects real power reversal. For steam turbine powered generators the relay operates when the power output falls below 3% of rated power output. A reverse power relay, having an operating setting of 0.2 to 2% of rated power for hydraulic sets and 15 and 50% settings for diesel sets and gas turbines respectively, is used.

7.11 UNIT GENERATOR-TRANSFORMER PROTECTION

Stepup transformer is usually installed adjacent to a large generator. It is possible to protect the transformer within the overall generator

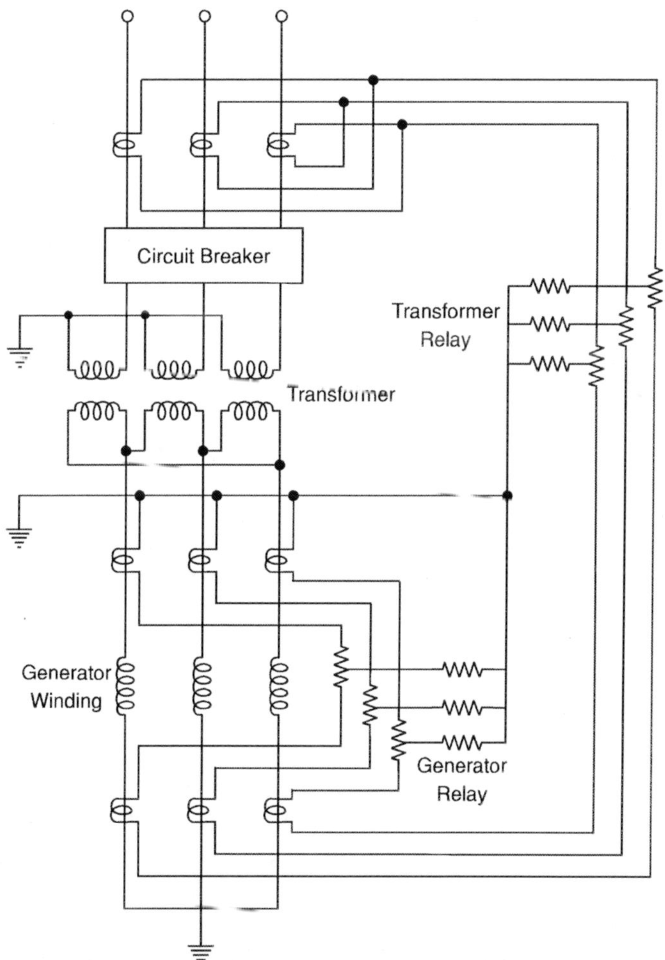

Fig 7.9 Unit generator-transformer protection

protection zone. Such type of protection is known as ***unit generator-transformer protection***. Figure 7.9 shows a unit generator-transformer protection scheme, The generator is directly connected to the delta connected primary winding of the power transformer. The secondary winding of the transformer is connected to the bus through a circuit breaker. An overall biased differential protection is provided to protect both the generator and transformer as one unit in addition to individual protection of transformer and generator.

Sometimes a balance current transformer can also be used to protect the transformer against earth faults.

7.12 AC MOTOR PROTECTION

Most of the electrical faults of motors are similar to those of generators. Abnormal conditions in motors can be due to

1. External conditions
2. Internal faults

Protection is to be provided against abnormal external conditions due to

- Unbalanced supply voltage and single phasing
- Overloading or stalling
- Undervoltage or overvoltage
- Reverse phase sequencing

Internal fault includes

- Stator faults
- Rotor faults (both electrical and mechanical)

Phase to phase, phase to ground, open circuiting are possible stator faults. Electrical faults on the rotor are ground faults or phase to phase faults in case of wound rotors. Core faults are also included in rotor electrical faults. Field failure may be caused in synchronous motors. Mechanical faults can be caused due to vibrations.

Unbalanced supply voltages including single phasing causes induction of double frequency currents in the rotor conductors and core leading to excessive rotor heating. Temperature sensing devices are embedded in rotor cores to detect such a situation. In addition, negative sequence detectors are placed in stator circuits to detect the flow of any negative sequence currents. A relay is set to operate if the negative sequence component is above a certain limit.

Overload, undervoltages, stalling or prolonged starting currents cause the current to increase above the rated current. This current causes overheating. Overcurrent relays and thermal detectors are employed for protection against such abnormal conditions.

In some applications like cranes, hoists, trams etc. reverse phase sequencing can be dangerous. Solid state phase reversal relays are normally used for protection against such conditions.

Almost all of the stator and rotor faults are taken care of by overcurrent thermal relays, negative sequence detectors etc. Large synchronous motors are protected using differential protection schemes and restricted earth fault protection schemes in stator and undercurrent relay and rotor ground fault protection in rotor.

EXERCISES

1. State various types of faults that can occur in an ac generator.

2. Explain the protection of a 3-phase star-connected generator by means of a percentage differential relay.

3. Explain the principle of operation of differential scheme. How do you apply it in the protection of an alternator.

4. Explain with the help of a line diagram the connections and functioning of differential relay for generator protection.

5. Explain the percentage differential protection of a star-connected generator.

6. What are the common types of alternator faults ? How is the generator stator protected against an interturn fault ?

7. Write a note on stator earth fault protection of an alternator.

8. How is rotor of an alternator protected against an earth fault ?

9. Explain the loss of excitation protection of alternators.

10. What are the common types of faults that occur in large motors.

11. What is unit generator-transformer protection ? Describe with the help of a diagram the unit generator-transformer protection scheme.

12. Explain with a neat diagram the protection of a generator-transformer unit in a power station.

❑❑❑

8

TRANSMISSION LINE AND BUSBAR PROTECTION

8.1 PILOT RELAYING

The term ***pilot relaying*** essentially means remote control of circuit breakers. That is, the decision to open or close a circuit breaker is made geographically remote (which may be many kilometres) from the circuit breaker location. Pilot relaying is the best type of line protection. It is used whenever high-speed protection is required for all types of short circuits and for any fault location. The main advantages of clearing of faults are as follows:

- Minimum damage to equipment
- Minimum transient stability problems
- Automatic reclosing

The term pilot implies that there is some type of channel (or medium) between the two ends of the line to be protected. Information on currents and voltages are transmitted through this channel. That is, these communication channels carry voltage and current information to the decision making location. Four basic channels are used.

1. Pilot-wire protection
2. Power line carrier protection

3. Microwave pilot
4. Fibre-optic link

Pilot relays provide only primary protection and, therefore, a separate set of distance relays is used for necessary backup protection.

8.2 PILOT-WIRE PROTECTION

Pilot-wore protection is used for short lines of length upto 10-15 km. In this method two wires are used to carry information signals from one end to the other end of the protected line. Pilot wires may be underground cables or a pair of overhead wires other than the power line conductors. The scheme is a unit protection and operates on the principle of differential protection. The schemes are similar to those used for generator and transformer differential protection. The length of the pilot wire imposes a large burden on the current transformers. It is uneconomical to use pilot wires for long lines. This type of protection does not provide backup protection.

Pilot-wire differential systems of feeder protection are of two types:

1. Circulating-current system
2. Balanced voltage system

8.2.1 Circulating-Current System

The principle of operation of a circulating-current system of differential protection of a single-phase feeder is shown in Fig. 8.1. Two equal-ratio current transformers are connected at each end of the protected line section. The secondary windings of the CTs are connected in series. Under load or fault conditions, the induced voltages in the secondaries of the CTs add together to produce a circulating current in the pilot wire circuit.

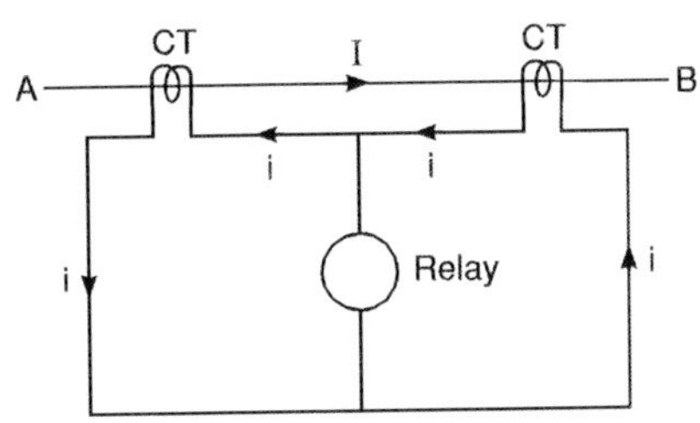

Fig. 8.1 Principle of circulating-current system of feeder protection

The relay is connected at the midpoint of the pilot circuit. It carries the difference between the two CT secondary currents. This difference current is zero since the secondary currents of the two CTs are identical. During an internal fault, there is a resultant current in the relay that operates to trip the faulted section from the section.

8.2.2 Balanced Voltage System

The principle of operation of a balanced-voltage system of differential protection of a feeder is shown in Fig. 8.2 for current transformers are connected at the ends of the protected line section. The secondary windings of the CTs are connected in ***series oppositions*** around the pilot loop. Under load or external fault conditions, there is no current in the relay that is connected in series with the pilot wires.

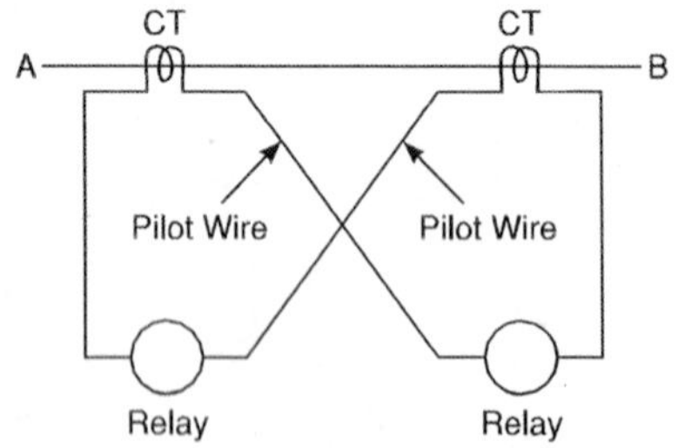

Fig. 8.2 Principle of balanced-voltage system of feeder protection

During an internal fault, the secondary voltages of the CTs are no longer equal and opposite. Therefore, the resultant voltage produces a current in the pilot wires and relay. The relay operates to disconnect the faulted circuit from the system.

8.3 POWER LINE CARRIER (PLC) PROTECTION

In power line carrier protection (PLC) system, the conductors of the power line to be protected are used for transmission of carrier signals. Therefore, this pilot is known as ***power line carrier pilot***. These pilots are widely used for the protection of extra high voltage (EHV) and ultra high voltage (UHV) power lines. The frequency range of carrier signals is 50 kHz to 700 kHz.

Carrier-current protection schemes are of two types :

1. Phase comparison schemes
2. Direction comparison schemes

In the *phase comparison scheme*, the phase angle of the current entering one end is compared with the phase angle of the current leaving the other end of the protected line section.

In the *directional comparison scheme*, the fault detecting relays compare the direction of power flow at the two ends of the protected line section. In case of an internal fault, the power flows inwards from both ends, and the line is tripped. During normal operating conditions and in case of external faults, power flows into the line at one end, and out of it at the other end, and tripping is not allowed.

8.4 PHASE COMPARISON CARRIER CURRENT PROTECTION

Figure 8.3 shows the schematic diagram of the phase comparison carrier current protection. The communication equipment is coupled to the power line at each end. The line trap (blocking filter) L_1 and C_1 is a parallel resonant circuit connected in series with the line conductor at each end of the protected line section. The line traps are tuned to the carrier frequency. They offer a low impedance at power frequency and high impedance to the carrier frequency signals. Thus, the carrier frequencies are prevented from entering the neighbouring sections and shorting out the channel. The signal is thus trapped between the ends of the line.

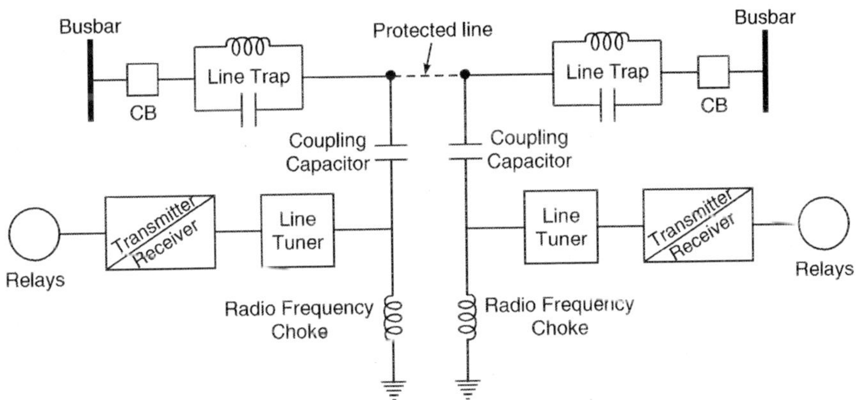

Fig. 8.3 Power line carrier for pilot relay system

Transmitters and receivers are provided at both the ends of the protected line. The transmitter and receiver are connected to the power line through a coupling capacitor. It offers high impedance to power-frequency current but low impedance to carrier-frequency current. Radio-frequency chokes are connected as shown in Fig. 8.3. The radio-frequency choke offers low impedance to power-frequency current and high impedance to carrier-frequency current. Therefore the transmitter and receiver are insulated from the power line and effectively grounded at power frequency. At carrier frequency the transmitter and receiver are connected to the power line and effectively insulated from the ground.

8.4.1 Modes of Coupling

There are three modes of coupling the carrier frequency signal to one or more of the conductors of the transmission line

1. Phase-to-ground coupling
2. Phase-to-phase coupling
3. Coupling to all three phases.

1. Phase-to-ground Coupling

This is simplest and most commonly used coupling on high voltage transmission lines. In this system one phase of the transmission line is used with ground as the return path. The return path is usually the overhead ground wire or the ground itself. This coupling is also known as 'line-to-ground', 'phase-to-ground' or 'single-phase coupling'. This system requires less coupling capacitors and tuners and wave traps.

2. Phase-to-phase Coupling

This coupling is between any two phases on the same 3-phase transmission line. This is also known as 'line-to-line' coupling. When the coupling is between adjacent lines on the same tower, the coupling is called 'interphase' coupling. Phase-to-phase coupling has an advantage that a ground on one of the phases does not affect the communication signal.

3. Coupling to All Three Phases.

For extra high voltage (EHV) lines, it is preferred to couple all the three phases. Such type of coupling provides greater channel capacity and less attenuation.

The three modes of coupling, in decreasing order of attenuation and increasing order of reliability and cost, are phase-to-ground coupling using centre phase, phase-to-phase coupling, phase-to-phase coupling between the outer phases, and coupling to all three phases.

Microwave channel is not affected by problems on the transmission line but is subject to atmospheric attenuation and distortion. The transmission length is limited to a line of sight path between antennas but can be increased through the use of intermediate repeater stations for increased cost and decreased reliability. A microwave link may operate upto 60 km without repeater stations.

8.5 MICROWAVE PILOT

A microwave pilot is a radio channel usually in the range of 450 to 10,000 MHz. Relaying information is transmitted by using beamed radio signals between the line-of-sight antennas located at terminals. A microwave channel has the advantage that considerable information can be transmitted in single beam. Microwave transmission is applicable only where there is a clear line of sight between stations.

8.6 FIBRE-OPTIC LINKS

A fibre-optic channel is an excellent channel for relaying. The channel capacity is unlimited with an optical-fibre cable. However, the use of fibre-optical cable is rarely justified just for protection, but with its large data transmission capacity, it is used for dispatching and telemetering. The fibre link is immune to electromagnetic interference and provides an excellent transmission quality.

There are several methods of stringing fibre-optic cable. They may be laid in cable trenches. Two of the most common methods are to embed the fibre cable within the aluminium conductors of the overhead ground wire or to wrap the fibre cable around one of the phase or ground conductors.

8.7 SELECTION OF A COMMUNICATION CHANNEL

The selection of a communication channel for protection is based on many factors such as number of terminals and the distance between then, reliability and cost. This is also to be decided whether the scheme is to be operated in a blocking mode or in a tripping mode.

A **blocking mode** is one in which the presence of a transmitted signal prevents the tripping of a circuit breaker. A **tripping mode** is one in which the signal initiates tripping a circuit breaker. There are different relay schemes for these modes. The decision to use a block or trip signal is based on the relationship between the power line and the communication chanel. If the communication medium is an integral part of the protected line section, such as power line carrier (PLC), blocking signal is used. In this cases an internal fault prevents the signal so that the trip signal is not received. The tripping mode is applied if a separate channel such as microwave, fiber-optic cable or a pilot cable is used. The power line has no effect on the transmitted signal during an internal fault when a separate transmission medium is used.

8.8 BUSBAR PROTECTION

Bus protection is used to protect switches, isolators, instrument transformers, circuit breakers and other bus equipment as well as the bus itself. The protection scheme should be fast, reliable, and secure. Usually differential relaying is used because of its great selectivity. A number of methods are used for bus protection. Two methods are described here.

8.9 DIFFERENTIAL PROTECTION OF A BUS USING LINEAR COUPLERS

Figure 8.4 shows the differential protection of a bus. In this scheme, discrimination between internal and external faults is made on the basis of the voltage magnitude that appears across the relay.

When an external fault takes place the voltage produced across the relay circuit is very small. This is due to the fact that the secondary circuit is only opposed, at the most, by the CT terminal resistance and internal secondary impedance of the CT in the faulted circuit when it is completely saturated. In case of an internal fault, the secondary current is opposed by the magnetizing impedance of the CTs and by the high impedance of the relay circuit. Thus, a large operating voltage is produced across the relay and it operates.

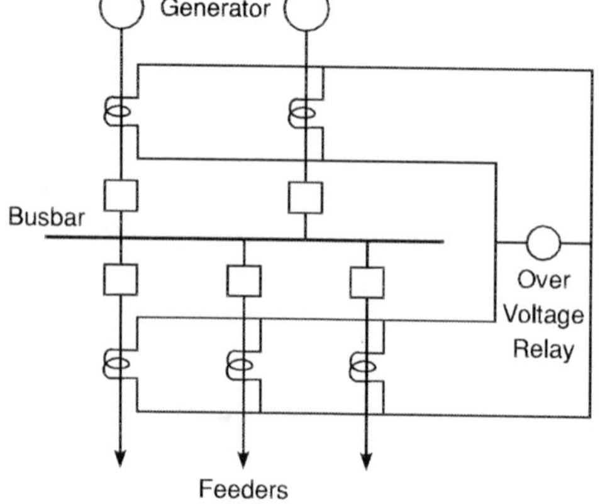

Fig. 8.4 Differential protection of a bus

In this scheme, false relay operation can be prevented by adjusting the minimum pick up of the relay well above the maximum voltage produced on an external fault. The problem of CT saturation can be eliminated at its source by using air core CTs called ***linear couplers***. These CTs are like bushing CTs but they have no iron in their core. The number of secondary turns of such a linear coupler is much greater than the secondary turns of a bushing CT. Contrasted with conventional CTs, linear couplers may be operated without damage with their secondaries open circuited.

8.10 DIFFERENETIAL PROTECTION OF A BUS USING HIGH-IMPEDANCE RELAY

Figure 8.5 shows a differential protection scheme of a bus using a high-impedance relay. In this scheme, conventional current transformers are used.

Bushing type transformers are employed because they have a low secondary impedance. The current transformers are loaded with a high impedance relay unit to nullify the unequal current transformer performance problem. A sensitive dc polarized relay R is operated through a full-wave rectifier. A tuning circuit consisting of a capacitance and an inductance is connected as shown in Fig. 8.5. The tuning circuit makes the relay responsive only to the fundamental component of the differential current of the CTs. This circuit reduces response to dc and harmonics. The impedance of this branch is around 3000 ohms, which means that CT secondaries and relay are subject to high voltages on a bus fault. A thyrite voltage limiting unit is connected in parallel with the relay to limit the voltage. An instantaneous overcurrent relay is connected in series with the combination and is set to operate at very high internal fault magnitudes. Its pickup is kept high to prevent operation on external faults. On internal faults, the voltage applied to the relay is high approaching the open-circuit voltage of the CT secondaries. On external faults the voltage will be approximately zero unless the CTs saturate.

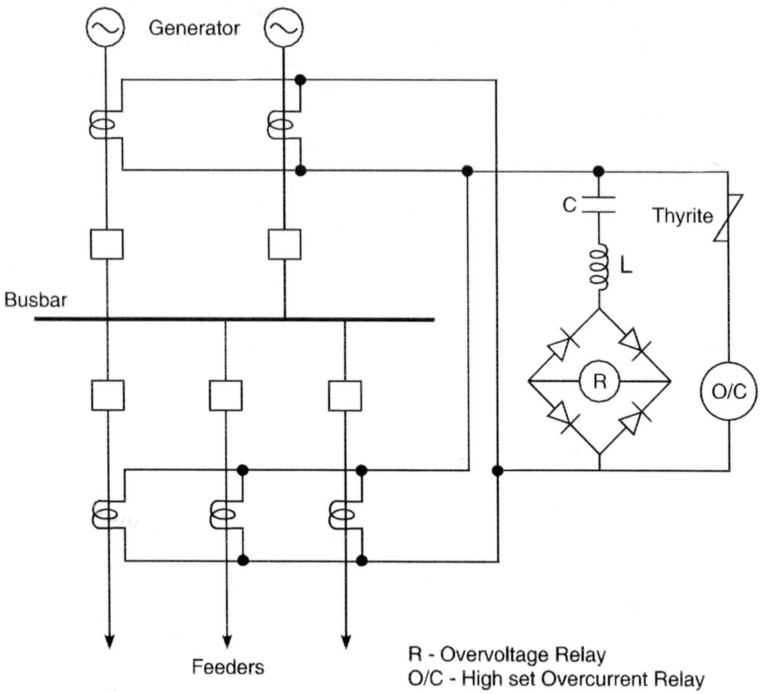

Fig. 8.5 Differential protection using high-impedance relay

8.11 AUTORECLOSING

The main objective of an electrical power system is to maintain continuity of power supply to consumers. In the event of a fault, the power supply is interrupted. About 80–90% of the faults on overhead lines are of transient nature. These faults are self clearing if the supply is interrupted by the circuit breakers for a short period. The arc path is allowed to de-ionize. The line is reenergized by reclosing circuit breakers. This is known as **reclosure**. The reclosure can be done manually or automatically. Automatic reclosing of a circuit breaker after its reopening is known as **autoreclosing**. By making a circuit breaker autoreclosing, transient faults do not reestablish themselves after restoration of supply. The time interval between the line removal and reclosure should be reduced keeping in mind that the line must remain deenergized for a certain minimum time in order that the line insulation should recover fully.

De-ionizing time for a line fault is defined as the time required for de-ionizing the are space after opening of the circuit breaker.

The fault should be cleared as fast as possible. It should be noted that the time required for fault removal is the sum of relay response time and the circuit breaker operating time.

The de-ionization time depends on the system voltage.

The typical de-ionization times for various system voltages are shown in Table 8.1.

Table 8.1 Typical De-energization Times for Autoreclosing

Line voltage in kV	66	132	220	400
Minimum de-energization time in seconds	0.10	0.17	0.28	0.50

The time required for de-ionization is equal to the dead time of the circuit breaker. The **dead time** of **circuit breaker** is defined as the time between the fault are being extinguishes and the circuit breaker contacts touching during subsequent reclosing.

8.12 DEFINITIONS

(a) Operating Time of Protective Relay

Time from the inception of the fault to the closing of the tripping contacts.

(b) Operating Time of Circuit Breaker

Time from the energizing of the trip coil until the fault arc is extinguished.

(c) Dead Time of Circuit Breaker or System

Time between the fault are being extinguished and the circuit breaker contacts remaking.

(d) Reclosing Time

Time taken by the circuit breaker to open and reclose the line. It is measured from the instant of energization of the trip circuit to the instant when the breaker contacts remake the circuit. This period is the sum of circuit breaker time and the system electrical dead time.

(e) Lockout of Circuit Breaker

A feature of autoreclosing scheme which prevents further automatic closing of the circuit breaker after the chosen sequence of reclosures have been unsuccessful. From this position, the circuit breaker must be reclosed manually.

(f) Number of Shots

The number of attempts at reclosing which an autoreclosing scheme will make before locking out on a permanent fault.

8.13 SINGLE-SHOT AND MULTI-SHOT AUTORECLOSING

Reclosure has different implications for low/medium (LV/MV) voltage and high/extra high voltage (HV/EHV) systems. In lower and medium voltage systems, the continuity of supply is the main consideration. In high voltage and extra high voltage systems, the main considerations are of stability and synchronizing.

On low and medium voltage lines, faults may be caused by external objects such as tree branches falling on the line. These branches are burnt at the first reclosure due to large amount of power in the arc. In a single-shot autoreclosing scheme only one reclosure is made. The tree branches may not be burnt clear at the first reclosure, and therefore, subsequent reclosure are required.

In **multi-shot autoreclosing scheme** more than one automatic reclosure, are made. In low and medium voltage systems, a maximum of three consecutive automatic reclosures are allowed. If the fault is not cleared after three reclosure, there is an automatic lockout of the circuit breaker and no more reclosures are allowed. The usual practice is to reclose the circuit breaker three times. The fourth reclosure, if required is done manually. If the fourth reclosure fails, there is clear indication of a permanent fault.

Only one reclosure is allowed in HV/EHV systems because the fault MVA is very large. In other words, **single-shot autoreclosing** is used in HV/EHV systems. High speed reclosure in HV/EHV systems improves the transient stability limit. The reclosure should be made as quickly as possible so that there should not be an appreciable drift in phase angle between the voltages at the two ends of the open line. This is due to the fact that as soon as there is an interruption in the system, the rotor angles of various generators start drifting apart. If they drift apart beyond a critical angle, the system becomes unstable. If however, the supply is restored before this critical time elapses, the system remains stable. If there are circuit breakers at both ends of an EHV, they should be tripped and reclosed simultaneously. For simultaneous tripping, pilot relaying is used for the remote control of circuit breakers.

8.14 SINGLE-PHASE (SINGLE-POLE) AUTORECLOSING

Majority of the line faults are single line-to-ground (LG) faults. In the event of an LG fault, the circuit breaker pole, corresponding to the faulty line is opened and the remaining two healthy phases continue to transfer power. Since most of the faults are transistory, this phase can be reclosed after it has been open for a predetermined time. The system should not be operated for long periods with one phase open. Therefore, provison should be made to trip the whole line if one phase remains open for a predetermined time. In a single-phase autoreclosing scheme, only the faulty phase pole of the circuit breaker is tripped and reclosed. For any multiphase fault, all the three phases are simultaneously tripped and reclosed. In single-phase autoreclosing, the three phases of the circuit are closed or opened simultaneously of each other. This is also called **single-pole switching** or **independent-pole operation**. In this scheme, each phase of the circuit breaker is provided with its own closing and tripping mechanism. The main disadvantage of single-phase autoreclosing is a longer deionization time. This is due to the fact that when a faulty phase is disconnected, a capacitive coupling between healthy phases and the faulty phase tends to maintain the arc. Interference with communication circuits may be produced. A complex relaying scheme is required in this mode of autoreclosing. Therefore, single-phase autoreclosing scheme is costlier than three-phase autoreclosing scheme.

8.15 THREE-PHASE AUTORECLOSING

In a three-phase autoreclosing schemes, all the three phases are tripped and reclosed in the event of a fault, irrespective of the types of fault. This scheme has a smaller de-ionizing time and simpler and less expensive than the single-phase autoreclosing scheme.

Autoreclosing is adopted only for overhead times. It is not used for underground cables or gas insulated cables.

EXERCISES

1. Explain the term "pilot" with reference to power line protection. What are the different types of pilots which are presently employed ? What are their fields of application ?

2. Explain the operating principles of pilot-wire protection ? Explain the principle of phase-comparison protection.

3. What type of pilot is used for the protection of extra high voltage and ultra high voltage transmission line ? What are the merits and demerits of carrier-current protection ?

4. Explain differential protection of a bus using linear couplers.

5. Explain differential protection of a bus using a high-impedance relay.

6. Discuss the merits and demerits of various pilot-wire relaying schemes for protecting transmission lines.

7. What is carrier-current protection ? Explain the phase comparison scheme of carrier current protection. Explain the utility of line trap unit and coupling capacitor unit.

8. Draw the schematic diagram of a carrier current protection scheme of lines. Also explain its working principle.

9. What is meant by reclosure ? What is the purpose of providing autoreclosing ?

10. Define the following terms :
(a) Dead time of circuit breaker
(b) Reclosing time
(c) Lockout of circuit breakers
(d) Number of shots

11. What are the advantages of autoreclosing ?
What is meant by the following terms :
(a) Single-shot autoreclosing
(b) Multishot autoreclosing
(c) Single-phase autoreclosing
(d) Three-phase autoreclosing
(e) Delayed autoreclosing

12. Distinguish between single-phase and three-phase autoreclosing. What are their relative merits and drawbacks ?

❑❑❑

9

SYSTEM NEUTRAL GROUNDING

9.1 INTRODUCTION

A *system neutral ground* is a connection to ground from the neutral point or points of a system or rotating machine or transformer. The neutral grounding is an important aspect of power system design because the performance of the system in terms of short circuits, stability, protection, etc., is greatly affected by the condition of neutral. A three-phase system can be operated in two possible ways:

1. With ungrounded neutral
2. With grounded neutral

9.2 UNGROUNDED NEUTRAL SYSTEM

In an ungrounded neutral system, the neutral is not connected to ground. In other words, the neutral is isolated from the ground. Therefore, this system is also known as ***isolated neutral system*** or ***free neutral system***. It is shown in Fig. 9.1.

The line conductors have distributed capacitances between one another and to ground. The former are delta connected while the latter are star connected as shown in Fig. 9.2. The delta connected capacitances have little effect on grounding characteristics of the system and therefore can be neglected.

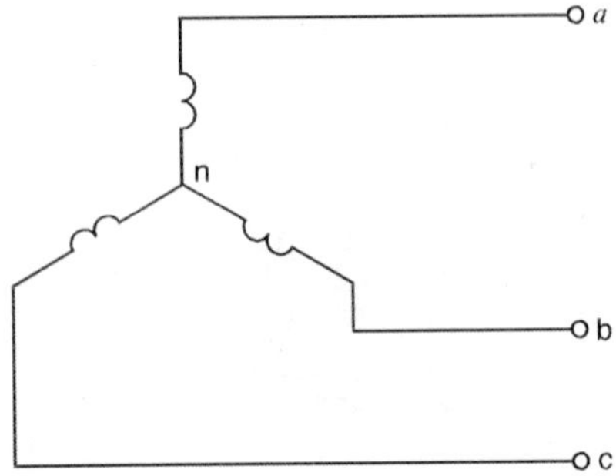

Fig. 9.1 Isolated neutral

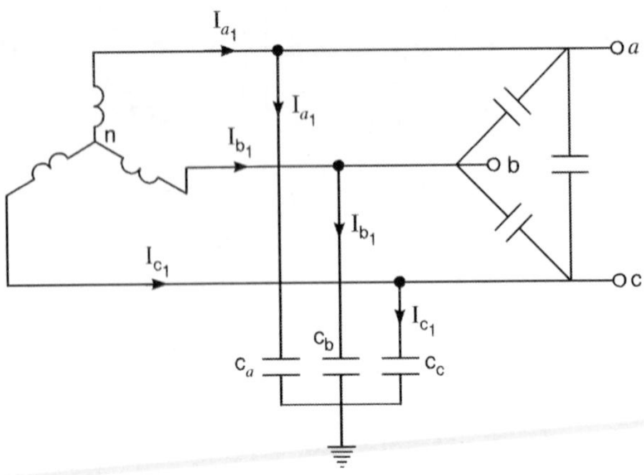

Fig. 9.2

A perfectly transposed line will have the same capacitance to ground. Thus, $C_a = C_b = C_c = C$. Under normal operating balanced conditions, the charging current of each phase is the same. They form a balanced set of currents spaced 120° apart. Therefore, their resultant is zero and no current flows to ground. The potential of the neutral is the same as the ground potential. Let the phase voltages (line-to-neutral voltages) be V_{an}, V_{bn} and V_{cn}. It is to be noted that the charging currents I_{a_1}, I_{b_1} and I_{c_1} lead their respective phase voltages by 90°. Figure 9.3 shows the phasor diagram under normal conditions.

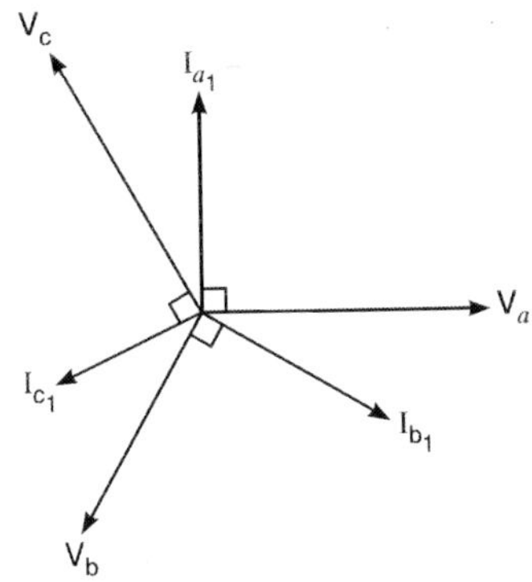

Fig. 9.3 Phasor diagram under normal conditions

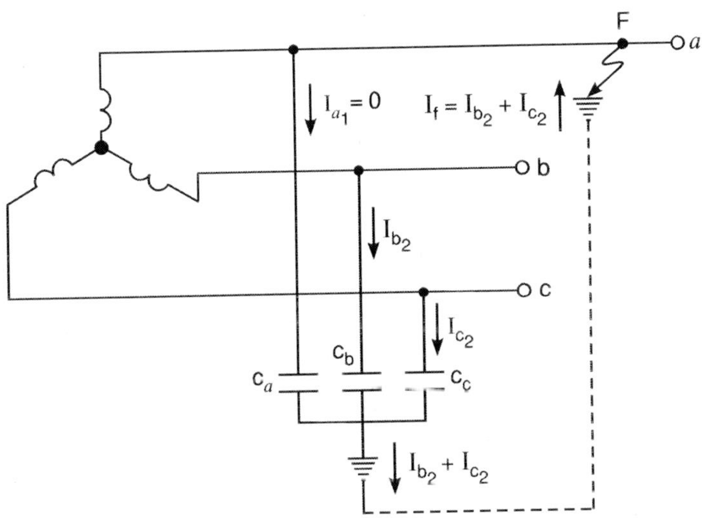

Fig. 9.4 Charging current during LG fault on line *a*

The magnitude of each charging current is given by

$$|I_{a_1}| = |I_{b_1}| = |I_{c_1}| = \frac{V_p}{X_c}$$

where,

V_p = phase voltage

X_c = capacitive reactance of the line to ground.

Now consider a line-to-ground fault in line *a* at a point F as shown in Fig. 9.4. As a result of this LG fault, the potential of phase *a* becomes equal to ground potential, thus short circuiting the capacitance of this line. Hence no charging current flows in this phase. That is $I_{a2} = 0$. It is to be noted that the potentials of two healthy phases b and c rise from phase values to the line values. The capacitance currents now become unbalanced and the fault current If is the phasor sum of the currents I_{b2} and I_{c2} which flow through the capacitances C_b and C_c under the potential differences of V_{ba} and V_{ca} respectively. These currents lead their respective voltages by 90°. The phasor diagram is shown in Fig. 9.5.

$$I_{b2} = \frac{V_{ba}}{X_{cb}} = \frac{\sqrt{3}V_p}{X_c}$$

$$I_{c2} = \frac{V_{ca}}{X_{cc}} = \frac{\sqrt{3}V_p}{X_c}$$

$$\therefore \quad I_{b2} = I_{c2}$$

$$I_f = I_{b2} + I_{c2}$$

From the phasor diagram in Fig 9.5

$$I_f = \sqrt{3}\, I_{b2} = \sqrt{3}\, I_{c2} = \frac{\sqrt{3} \times \sqrt{3}\, V_p}{X_c} = \frac{3V_p}{X_c}$$

$$\therefore \quad I_f = 3I_{b_1} = 3I_{c_1}$$

Thus, we arrive at the following conclusions in the event of a single line-to-ground fault on a system with isolated neutral:

1. The potential of the faulted phase becomes equal to the ground potential while the voltages of the remaining two healthy phases rise from their normal phase voltages (line-to-neutral voltages) to full line-to-line voltages.

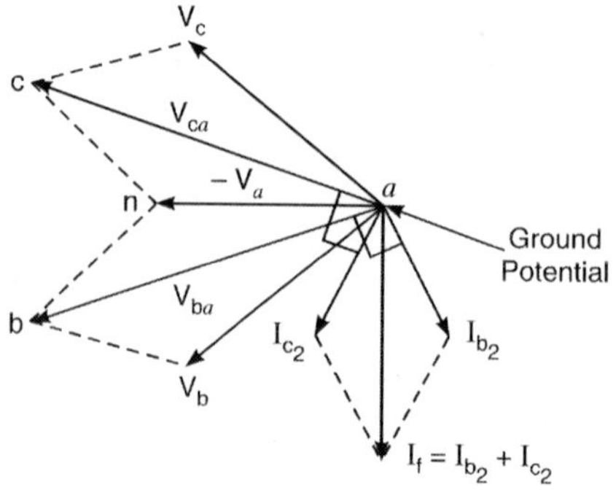

Fig 9.5 Phasor diagram with LG fault on line a

2. The charging current (capacitance current) in the two healthy phases increases to $\sqrt{3}$ times their normal values.
3. The fault current becomes three times the normal per phase charging current under balanced conditions.

9.3 ARCING GROUNDS

The insulation of all equipment connected to the lines is subjected to the high voltage. If it exists for a very short duration, the insulation may be able to withstand it, otherwise it may be damaged. For operating the protective devices, it is necessary that the current should be sufficient in magnitude. However, in case of a single line-to-ground fault on an isolated neutral system, the resultant capacitive current is usually large enough to operate the protective device. Further, a current of over 4 or 5 A flowing through the fault may give rise to an arc in the ionized part of the fault. With the formation of the arc, the voltage across it becomes zero and, therefore, the arc is extinguished. The potential of the faulty conductor is restored and the formation of second arc takes place. The phenomenon of intermittent arcing is called the ***arcing ground***. The successive extinction and reignition of the charging current flowing in the arc may increase the potential of the other two healthy conductors appreciably due to setting up of high-frequency oscillations. These high-frequency oscillations are superimposed on the system and produce surge voltages as high as six times the normal value. The overvoltages in healthy conductors may damage the insulation at some other point of the system.

9.4 ADVANTAGES OF NEUTRAL GROUNDING

Because of the problems associated with ungrounded neutral systems, the neutrals are grounded in most of the modern high-voltage systems. Some of the advantages of neutral grounding are as follows:

1. Voltages of phases are limited to the line-to-ground voltages.
2. Surge voltages due to arcing grounds are eliminated.
3. The overvoltages due to lightning are discharged to ground.
4. The induced static charges do not produce any disturbance since they are conducted to ground.
5. The ground relays can be used to protect against the ground faults.
6. It provides greater safety to personnel and equipment.
7. It provides improved service reliability.
8. Operating and maintenance expenditure is reduced.

9.5 METHODS OF NEUTRAL GROUNDING

The methods commonly used for grounding the system neutral are:

1. Solid grounding (or effective grounding)
2. Resistance grounding
3. Reactance grounding
4. Peterson-coil-grounding (or resonant grounding).

The selection of the type of grounding depends on the size of the unit, system voltage and protection scheme to be used.

9.6 SOLID GROUNDING

The term effectively grounded is now used instead of the old term "solidly grounded". A power system is said to be effectively grounded when the neutral of a generator, power transformer, or grounding transformer are directly connected to ground through a conductor of negligible resistance and negligible reactance as shown in Fig. 9.6. A system or a portion of a system is defined to be effectively grounded when $R_0 \leq X_1$ and $X_0 \leq 3X_1$, such relationships exists at any point in the system for any condition of operation and for any amount of generator capacity.

Here,

R_0 = zero-sequence resistance

X_0 = zero-sequence reactance

X_1 = positive-sequence reactance.

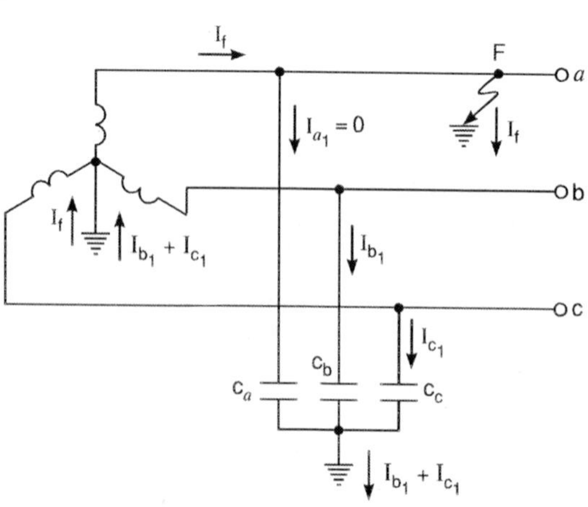

Fig. 9.6 Solidly grounded system

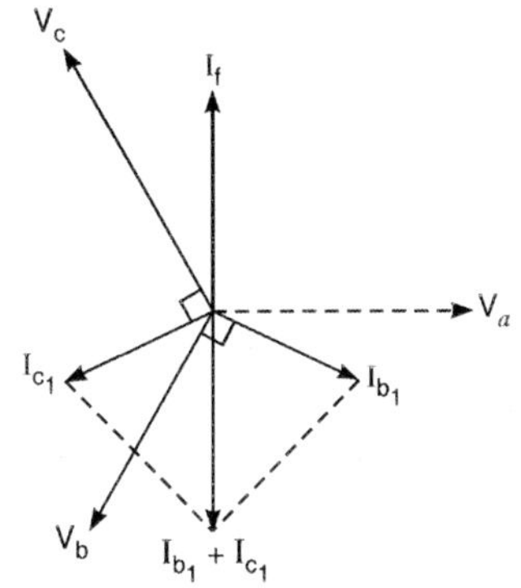

Fig 9.7 Phasor diagram of a solidly grounded system.

Consider a line-to-ground fault in line a at point F as shown in Fig. 9.6. As a result of this fault the line-to-ground voltage of phase a becomes zero. However, the remaining two phases b and c will still have the same voltages as before as shown in Fig. 9.7.

It should be noted that in this system, in addition to charging currents, the power source also feeds the fault current If. As the generator or transformer have their own reactances in series with the neutral circuit, solid grounding does not make a zero-impedance circuit. If the impedance of the generator is too low, solid grounding of the generator without any external impedance may cause the single line-to-ground fault current from the generator to exceed the maximum three-phase fault current which the generator can deliver, and this may exceed the short-circuit current for which its windings are braced. If the reactance of the generator or transformer is very large, then also the purpose of grounding is defeated. For solidly grounded systems, it is necessary that the ground fault current should not exceed 80% of the 3-phase fault current to prevent the production of surge voltages.

9.7 RESISTANCE GROUNDING

In a resistance-grounded system, the system neutral is connected to ground through one or more resistors as shown in Fig. 9.8. Resistance grounding limits the ground fault currents. A system that is properly grounded by resistance is not subject to dangerous transient overvoltages. Resistance grounding reduces the arcing ground hazards and permits ground-fault protection.

The value of resistance to be used in the neutral to be grounded should neither be very low nor very high. A very low resistance, makes the system similar to the solidity grounded system. If the grounding resistance is high, the system conditions becomes similar to an isolated (ungrounded) neutral system. The value of resistance is chosen such that the ground-fault current is limited but still sufficient ground current flows to permit the operation of ground fault protection. In general, the ground-fault current may be limited to 5% to 20% of that which would occur with a three-phase fault.

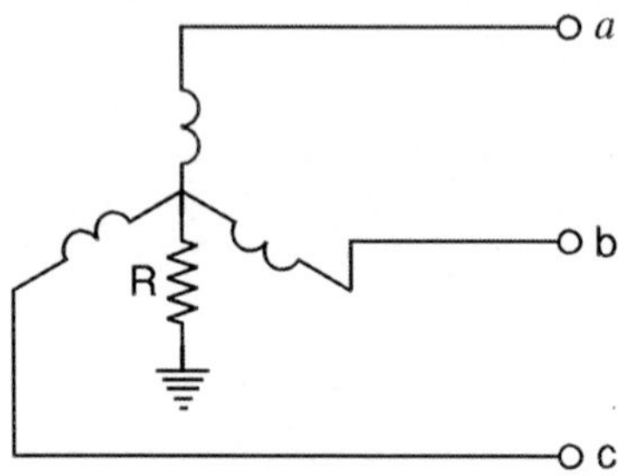

Fig. 9.8 Resistance grounding

9.8 REACTANCE GROUNDING

In this system a reactance is inserted between the neutral and ground to limit the fault current as shown in Fig. 9.9. In order to minimize transient overvoltages, the ground-fault current of a reactance grounded system should not be less than 25% of the 3-phase fault current. This is considerably more than the minimum current desirable in resistance-grounded systems.

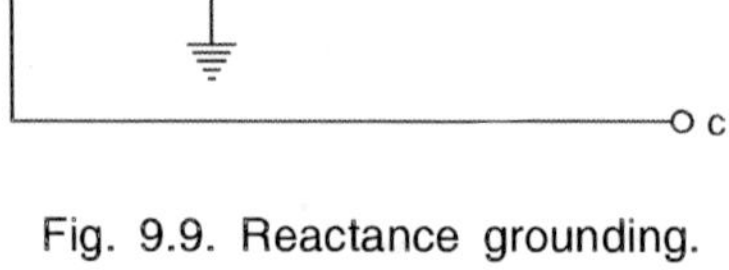

Fig. 9.9. Reactance grounding.

9.9 PETERSON-COIL GROUNDING

Peterson coil is an iron-cored reactor connected between neutral and ground. It is provided with tappings so that it can be tuned with the capacitance of the system. The reactance is selected so that the current through the reactor is equal to the small line-charging current which would flow into the line-to-ground fault if the system were operated with the neutral ungrounded. The reactance of the reactor balances the system ground capacitance so that the resultant ground-fault current is practically zero. This method is also known as resonant grounding.

Consider an *LG* fault in line a at a point F as shown in Fig. 9.10. Consequently the line-to-ground voltage of phase a becomes zero and $I_a = 0$. The voltages of the two healthy phases b and c increase from phase values to line values. The currents through the capacitances C_b and C_c becomes I_{b2} and I_{c2}. The current I_{b2} leads V_{ba} by 90° and the current I_{c2} leads V_{ca} by 90°.

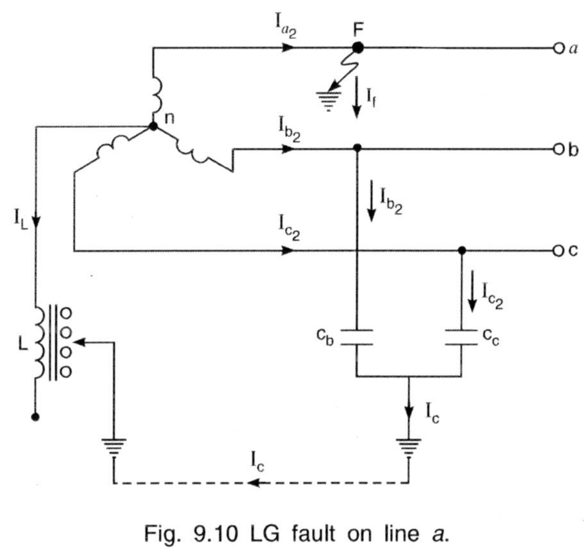

Fig. 9.10 LG fault on line *a*.

The resultant of I_{b2} and I_{c2} is I_c. The phasor diagram is shown in Fig. 9.11.

$$I_{b2} = \frac{V_{ba}}{X_{cb}} = \frac{\sqrt{3}V_p}{X_c}, \quad I_{c2} = \frac{V_{ca}}{X_{cc}} = \frac{\sqrt{3}\,V_p}{X_c}$$

$$I_{b2} = I_{c2}$$

$$\therefore \quad I_f = I_{b2} + I_{c2}$$

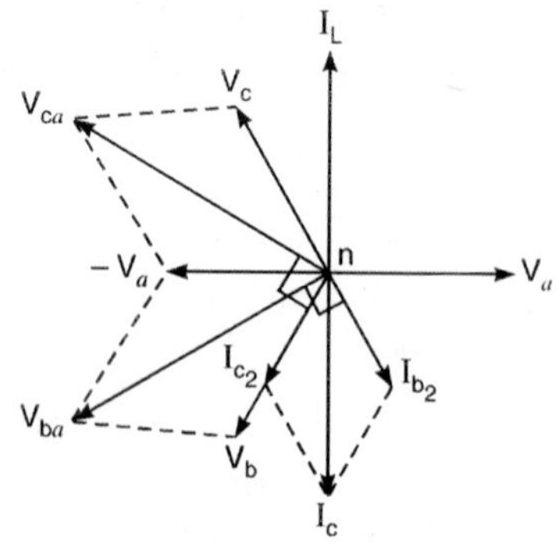

Fig. 9.11 Phasor diagram with LG fault on line a

From the phasor diagram in Fig 9.11

$$I_f = \sqrt{3}\, I_{b2} = \sqrt{3}\, I_{c2} \frac{\sqrt{3} \times \sqrt{3}\, V_p}{X_c} = \frac{3V_p}{X_c}$$

The current I_L through the inductances L of the Peterson coil is given by

$$|I_L| = \left| \frac{V_{na}}{\omega L} \right| = \frac{V_p}{\omega L}$$

But V_{na} is equal to $-V_a$. The current lags behind V_{na} by 90°. The current I_c in the phase opposition I_L. Hence if I_L is equal to I_c, there will be no current through and there will be no tendency of the arcing grounds to occur. With the use of Peterson coil, arc current is reduced to such a small value that it is usually self extinguishing. Therefore, Peterson coil is also known as ***ground fault neutralizer*** or ***arc suppression coil***.

The inductance value for the Peterson coil can be calculated as follows:

For no current through the ground fault,

$$|I_L| = |I_c|$$

$$\frac{V_p}{\omega L} = \frac{3V_p}{X_c} = 3V_p \omega C$$

$$L = \frac{1}{3\omega^2 C} \quad \text{...(9.9.1)}$$

Equation (9.9.1) can also be written as

$$\omega = \frac{1}{\sqrt{(3C)L}}$$

But this relationship, is also the condition for resonance at the supply frequency ω rad/s, when a capacitance of 3C farad is connected in series with an inductance of L henry. Hence the required condition for the quenching of the arc at the fault point F is that the reactance (L shall be in series resonance with the capacitance 3C farad, where C farad is the capacitance of each line to ground.

It is seen from Eq. (9.9.1) that the Peterson coil depends for its operation, on such an adjustment of its reactance that

$$\omega L = \frac{1}{\omega \times (3C)} \qquad ...(9.9.2)$$

In other words, this condition states that the inductance must be "tuned" to the capacitance if the Peterson coil is to fulfill its function.

9.9.1 Construction of Peterson Coil

The construction of Peterson coil is similar to that of an oil immersed transformer. It consists of a winding on a magnetic core. The coil is provided with tapping so that its inductance can be adjusted or tuned to such a value that the current passing through it is equal and opposite to the resultant current passing through the ground capacitances of the healthy lines in the event of a single line-to-ground fault.

9.9.2 Rating of Peterson Coil

Peterson coil may be rated either for a short-time duty of about 5 minutes, or it is designed to carry its rated current continuously. A short-time rated coil is provided with an automatic circuit breaker, which by the action of a relay shorts the coil after a definite time and connects the neutral directly to ground. A short-time rated coil clears transient faults without supply interruption. Sustained faults are cleared by isolating the faulty section. A continuously rated coil allows a fault to remain in the system till it is located and removed.

9.9.3 Advantages of Peterson-coil grounding

1. The Peterson-coil grounding is an effective method of clearing both transient faults due to lightning and sustained single line-to-ground faults.

2. There is no tendency of arcing grounds to occur and the arcs are usually self extinguishing.
3. Voltage drops to single line-to-ground faults are minimized.

9.9.4 Limitations of Peterson-coil grounding

1. There is a need for retuning after any network modification.
2. The lines should be transposed.
3. There is an increase in corona and radio interference (RI) in the event of a double line-to-ground fault.

9.10 GROUNDING TRANSFORMER

If the system neutral is not available for some reason, for example, when a system is delta connected, the grounding can be done as follows:

(a) By using a zigzag grounding transformer

(b) By using a delta star grounding transformer.

(a) Zigzag Transformer

The zigzag transformer has no secondary winding. It is a three-limbed transformer. Each limb has got two identical windings. One set of windings is connected in star to provide the neutral point. The other ends of this set of windings are connected to the second set of windings as shown in Fig. 9.12. The directions of current in the two windings on each limb are opposite to each other.

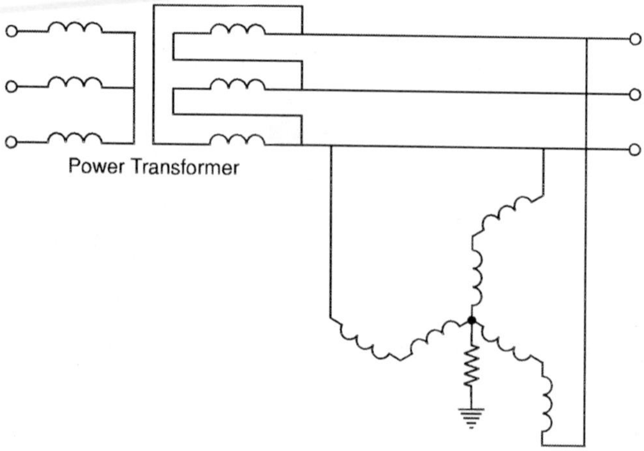

Fig. 9.12 Zig-zag transformer for neutral grounding

Under normal operating conditions the total flux in each limb is negligibly small. Therefore, the transformer draws very small magnetizing current. Under fault conditions the impedance of the grounding transformer is very low. In order to limit the fault current, a resistor is connected in series with the neutral grounding connection as shown in Fig. 9.12.

(b) Delta-Star Grounding Transformer

In case of delta-star grounding transformer, (Fig. 9.13) the delta side is closed to provide a path for zero-sequence current. The star winding must be of the same voltage rating as the circuit that is to be grounded, whereas the delta voltage rating can be chosen to be any standard voltage level.

The zigzag grounding transformer is designed for a short-time kVA rating equal to the rated phase voltage multiplied by the rated neutral current. The grounding transformer is designed to carry the rated current for very short time of 10 seconds or a maximum of one minute. The grounding transformer is therefore much smaller than an ordinary 3-phase transformer of the same rated kVA capacity.

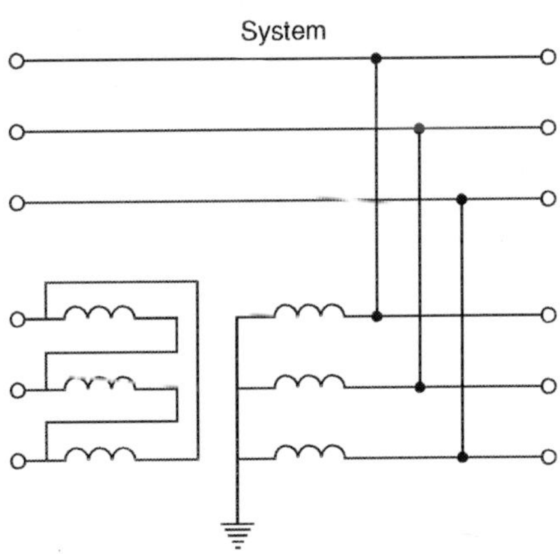

Fig. 9.13 Delta-star transformer grounding

9.11 CHOICE OF GROUNDING

The selection of the type of grounding depends on the type of the system and its voltage levels. The following considerations are made:

1. Transient overvoltages developed.

2. Magnitude of ground-fault current as a percentage of 3-phase fault current.
3. Dip in line voltage due to fault conditions.

Generally, solid grounding is used for low-voltage system (upto 600 V). For voltages up 11 kV resistance grounding is used. Arc suppression coils are best suited for high-voltage over head lines.

Example 9.1. A 50 Hz overhead line has line-to-earth capacitance of 1.2 μF. It is decided to use an earth fault neutralizer. Determine the reactance to neutralize the capacitance of (a) 100% of the length of the line, (b) 90% of the length of the line, and (c) 80% of the length of the line.

Solution : (a) The inductive reactance to neutralize 100% of the capacitance is given by

$$\omega L = \frac{1}{3\omega C} = \frac{1}{3 \times 2\pi \times 50 \times 1.2 \times .10^{-6}} = 884.19\ \Omega$$

(b) The capacitance of 90% of the length of line

$$= 0.90 \times 1.2 = 1.08\ \mu\text{F}$$

The inductive reactance to neutralize 90% capacitance

$$= \frac{1}{3\omega \times 0.9C} = \frac{1}{3 \times 2\pi \times 50 \times 0.9 \times 1.2 \times .10^{-6}} = 982.43\ \Omega$$

(c) The inductive reactance to neutralize 80% capacitance

$$= \frac{884.19}{0.80} = 1105.24\Omega$$

Example 9.2. A 132 kV, 3-phase, 50 Hz transmission line 80 km long has capacitance to earth of each line equal to 0.00914 μF per km. Determine the inductance and kVA rating of the arc suppression coil suitable for this line.

Solution : $C = 0.00914\ \mu\text{F/km} = 0.00914 \times 80 \times 10^{-6} = 0.7312 \times 10^{-6}\ \text{F}$

$$L = \frac{1}{3\omega^2 C} = \frac{1}{3\,(2\pi f)^2 C}$$

$$= \frac{1}{3\,(2\pi \times 50)^2 \times 0.7312 \times 10^{-6}} = 4.619\ \text{H}$$

$$I_L = \frac{V_{ph}}{X_L};\quad V_{ph} = \frac{V_L}{\sqrt{3}} = \frac{132 \times 10^3}{\sqrt{3}}\ \text{V}$$

$$X_L = 2\pi f L$$

$$= 2\pi \times 50 \times 4.619$$

$$\therefore \quad I_L = \frac{132 \times 10^3}{\sqrt{3} \times 2\pi \times 50 \times 4.619}$$

$$= 52.5188 \text{ A}$$

Rating of the a.c. suppression coil

$$S_{pc} = V_{ph} I_L = \frac{132 \times 10^3}{\sqrt{3}} 52.5188$$

$$= 4002.59 \times 10^3 \text{ VA} = 4002.59 \text{ kVA}$$

Alternative Method

Rating of the arc suppression coil

$$S_{pc} = V_{ph} I_L \ ; \quad I_L = \frac{V_{ph}}{X_L} \ ; \ X_L = \omega L = \frac{1}{3\omega C}$$

$$\therefore \quad S_{pc} = V_{ph}^2 \times 3\omega C = \left(\frac{V_L}{\sqrt{3}}\right)^2 . 3\omega C = V_L^2 \, (2\pi f) C$$

$$= (132 \times 10^3)^2 \times 2\pi \times 50 \times 0.7312 \times 10^{-6} \text{ VA}$$

$$= 4002.52 \times 10^3 \text{ VA} = 4002.52 \text{ kVA}$$

EXERCISES

1. Explain the phenomenon of arcing grounds. How does neutral grounding eliminate arcing grounds.
2. Explain the advantages of grounding power system neutrals.
3. What ate the various methods of neutral grounding ? Compare their performance with respect to (a) protective relaying, (b) fault levels, (c) stability, and (d) voltage levels of power system.
4. Explain the following statement
   In the event of a SLG fault on a system with isolated neutral the voltages of the remaining two healthy phases rise from their normal phase voltages to full line-to line voltages.
5. Explain with the help of circuit and phasor diagrams the function of a Peterson coil in a 3-phase system.
6. Derive an expression for the inductance of the Peterson coil in terms of the capacitance of the protected line.
7. A 33 kV, 3-phase, 50 Hz, 60 km overhead line has a capacitance to earth of each line equal to 0.01 μF/km. Determine the inductance and kVA rating of the arc suppression coil.

8. A 50 Hz transmission line has a capacitance of 0.1 μF per phase. Determine the inductance of Peterson coil to neutralize the effect of capacitance of (a) complete length of line, (b) 95% of the length of line, (c) 85% of the length of the line.

9. A 132 kV, 50 Hz, 3-phase, 75 km long transmission line has a capacitance of 0.015 μF/km per phase. Determine the inductive reactance and MVA rating of the arc suppression coil suitable for the line to eliminate arcing ground phenomenon.

10. A 132 kV, 50 Hz, 3-phase, 100 km long transmission line has a capacitance of 0.012 μF/km per phase. Determine the inductance and MVA rating of the arc suppression coil suitable for the line.

ANSWERS

7. 5.629 H ; 205.27 kVA

8. (a) 33.77 H (b) 35.55 H (c) 39.73 H

9. 943.14 Ω ; 6.158 MVA

10. 2.814 H ; 6.5687 MVA

❑❑❑

10

OVERVOLTAGE PROECTION

10.1 INTRODUCTION

When the current, voltage, or power changes in a circuit, there is a transient state between two steady state conditions. The transient behaviour of transmission lines differs from its steady-state behaviour. The transient period, although of a very short duration of the order of a few microseconds, is sufficient to cause much disturbance in the line. The transients are set up in the transmission lines mainly due to switching, faults and lightning. A ***surge*** is a movement of charge along the conductor. Such surges are characterized by a sudden very steep rise in voltage (***the surge front***) followed by a gradual decay in voltage (***the surge tail***). The surges produced on the line due to lightning are of particular interest due to their magnitudes and different waveshapes. The surges reach the terminal apparatus such as cable boxes, transformers or switchgear, and may damage them if they are not properly protected. As the waves travel along the line their waveshapes and magnitudes are also modified.

If there is any change in current, voltage, or a constant parameter at any point in line, voltage surges are set up. These voltages travel along the line at a very high speed. They are accompanied by current waves. The current and voltage waves travelling in forward direction have got the same sign. If they travel in backward direction, they are opposite in sign. To distinguish between the two waves travelling in opposite directions, use is made of suffix f to represent a wave travelling from the source towards the load and suffix r to represent the wave in opposite direction. Thus, e_f and i_f are used to represent the voltage and current waves travelling in the forward direction. And e_r and i_r the waves in the opposite direction.The waves traveling from the source towards the load are known as ***incident waves*** and those from load to the source are called ***reflected waves***.

These waves may superpose each others and the resultant wave at any point is the sum of incident and reflected waves there. Thus, if e_R and i_R denote the resultant voltage and current respectively at any point on the line.

$$e_R = e_f + e_r \qquad ...(10.1.1)$$

and

$$i_R = i_f + i_r \qquad ...(10.1.2)$$

Fig. 10.1 illustrates the two waves travelling in opposite directions.

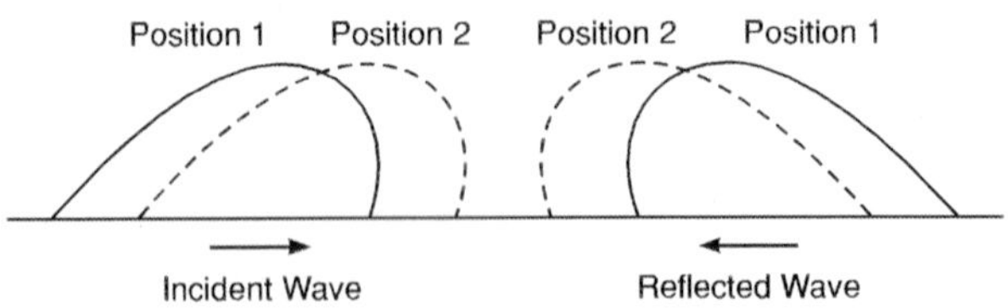

Fig. 10.1 Two waves travelling in opposite direction

10.2 REFLECTION OF WAVES

If the uniformity of the line is disturbed at a point, the surge impedance changes there. There will be a partial reflection and partial transmission of the wave when it reaches the point of discontinuity, that is, the incident wave gives rise to two waves, one transmitted forward and the other reflected back. The discontinuity in the line may occur as a consequence of the change in the characteristics of the circuit, for example, junction of an overhead line and a cable, the end of the line, forked line, the junction of the line with terminal apparatus, etc.

Consider a general case of a line which is terminated in an impedance Z_t , which is a pure non-inductive resistance. By Ohm's law,

$$\frac{\text{the resultant voltage at the receiving end}}{\text{the resultant current at the receiving end}} = \text{impedance at the receiving end}$$

$$\frac{e_R}{i_R} = Z_t \qquad ...(10.2.1)$$

Substituting the values of e_R and i_R from Eq. (10.1.1), (10.1.2), (10.2.1)

$$\frac{e_f + e_r}{i_f + i_r} = Z_t \qquad ...(10.2.2)$$

For incident wave,

$$e_f = i_f Z_0 \qquad ...(10.2.3)$$

And for reflected wave

$$e_r = -i_r Z_0 \qquad ...(10.2.4)$$

The values of currents obtained from these equations when substituted in the expression for Z_t give

$$\frac{e_f + e_r}{\frac{e_f}{Z_0} - \frac{e_r}{Z_0}} = Z_t \qquad ...(10.2.5)$$

Normally, the value of the incident voltage being known and therefore, the reflected voltage e_r is given by

$$e_r = e_f \frac{Z_t - Z_0}{Z_t + Z_0} \qquad ...(10.2.6)$$

The voltage at the end of the line is

$$e_R = e_f + e_r = 2e_f \frac{Z_t}{Z_t + Z_0}$$

The current through the impedance Z_t at the end of the line is

$$i_R = \frac{e_R}{Z_t} = \frac{2e_f}{Z_t + Z_0}$$

The ratio of the reflected voltage to the incident voltage is called the ***reflection coefficient*** or the ***reflection factor*** for voltage. It is denoted by ρ_r.

$$\rho_r \triangleq \frac{e_r}{e_f} = \frac{Z_t - Z_0}{Z_t + Z_0} \qquad ...(10.2.7)$$

This implies that a voltage wave is reflected at the termination and its amplitude becomes ρ_r times the incident wave. The ratio of the reflected current to the incident current is

$$\rho_i \overset{\Delta}{=} \frac{i_r}{i_f} = \frac{-e_r / Z_0}{e_f / Z_0} = -\frac{e_r}{e_f} = -\rho_r \qquad ...(10.2.8)$$

$\therefore$

$$\rho_r = \rho_i = \rho \text{ (say)} \qquad ...(10.2.9)$$

The physical meaning of the negative reflection coefficient needs an explanation. The current is assumed to be positive in the direction of the incident voltage wave, i.e., from the sending end to the receiving end. It is negative in the direction of the reflected voltage wave. The reflected voltage is positive. Therefore, a positive voltage at the receiving end will cause a current to flow from the receiving end to the sending end. But it has been assumed that the current is taken positive in the direction from the sending end to the receiving end. The current in the direction from the receiving end to the sending end should, therefore, be taken negative.

10.3 TRANSMISSION OR REFRACTION OF WAVES

It has been noticed earlier that there is a partial reflection and partial transmission of a wave at a point of discontinuity. The ratio of voltage at the termination to the incident voltage is called the ***transmission coefficient*** or ***refraction coefficient***. It is denoted by τ.

$$\frac{e_R}{e_f} \overset{\Delta}{=} \tau \qquad ...(10.3.1)$$

$$\frac{e_R}{e_f} - 1 = \tau - 1\,; \quad \frac{e_R - e_f}{e_f} = \tau - 1$$

i.e.,

$$\rho = \tau - 1 \qquad ...(10.3.2)$$

The values of reflected and resultant current and voltage for their incident value of i_f and e_f can be conveniently written as:

Reflected voltage,

$$e_r = \rho\, e_f \qquad ...(10.3.3)$$

Reflected current,

$$i_r = -\rho\, i_f \qquad ...(10.3.4)$$

Resultant voltage at termination,

$$e_R = e_f \frac{2Z_t}{Z_t + Z_0} = \tau\, e_f \qquad ...(10.3.5)$$

Resultant current at termination,

$$i_R = i_f \frac{2Z_0}{Z_t + Z_0} \quad ...(10.3.6)$$

In the above discussion the value of the terminating impedance is assumed to be a general one. In practice, it can have any value. But here only those cases will be considered which are of particular interest. They are as follows:

(a) When $Z_t = \infty$ i.e., the line is open circuited.

(b) When $Z_t = 0$, i.e., the line is short circuited.

(c) When $Z_t = Z_0$, i.e., the line is terminated in its characteristic impedance.

10.4 OPEN CIRCUITED LINE

When the line is open at the receiving end, Z_t becomes infinite and

$$\rho = \frac{Z_t - Z_0}{Z_t + Z_0} = \frac{1 - \frac{Z_0}{Z_t}}{1 + \frac{Z_0}{Z_t}} = +1 \quad ...(10.4.1)$$

Also, $\tau = 2$

The amplitude of the reflected wave is equal to that of the incident wave, the two waves being of the same sign. The resultant voltage at the receiving end

$$e_R = e_f + e_r = e_f + e_f = 2e_f \quad ...(10.4.2)$$

Also, for an open circuit,

$$i_R = 0$$

i.e.,

$$i_f + i_r = 0$$

$$i_r = -i_f \quad ...(10.4.3)$$

The direction of the reflected current is opposite to that of the incident current. The voltage at the receiving end of an open circuited line thus becomes equal to two times the incident voltage, and the current there is zero.

10.5 SHORT CIRCUITED LINE

For a line which is short circuited at the receiving end, $Z_t = 0$. Therefore, the voltage there will be zero, i.e.,

$$e_f + e_R = e_r = 0$$

Thus, $e_f = -e_r$ and $\rho = -1,\ \tau = 0$

Also, $i_f = i_r$ and $i_R = i_f + i_r = 2i_f$

These equations show that at the receiving point of the reflected voltage will be equal to incident voltage but of opposition sign. The reflected current is also equal to incident current but of the same sign. The, resultant voltage is zero at the termination, while the resultant current is double there.

10.6 LINE TERMINATED IN Z_0

If the line is terminated in its characteristic impedance Z_0, then $Z_t = Z_0$

$$\rho = \frac{Z_r - Z_0}{Z_t + Z_0} = 0$$

$$\tau = 1,\ e_r = 0,\ i_r = 0$$

which shows that there will be no reflected wave, i.e., no reflection occurs. The incident waves of voltage and current will be entirely absorbed by the load. Such a line with the receiving-end impedance equal to its characteristic impedance is said to be ***correctly terminated*** or ***matched.*** It is also called ***a flat line*** on ***infinite line.***

10.7 JUNCTION OF TWO DISSIMILAR LINES

If surge traveling along the line *A* of surge impedance Z_1 meets a junction of a line *B* of surge impedance Z_2, there will be a partial reflection of wave at the junction and partially the surge will be transmitted into the line *B*.

Let e_f, i_f = incident waves of the voltage and current

e_r, i_r = reflected waves of the voltage and current

e_t, i_t = voltage and current waves transmitted into the line.

Then
$$\left.\begin{aligned} i_f &= \frac{e_f}{Z_1} \\ i_t &= \frac{e_t}{Z_2} \\ i_r &= -\frac{e_r}{Z_1} \end{aligned}\right\} \qquad ...(10.7.1)$$

Since the total current in the two lines at the junction is the same.

$$i_t = i_f + i_r \quad ...(10.7.2)$$

Also the total voltage at the junction will be equal to the algebric sum of the incident and reflected voltages, i.e.,

$$e_t = e_f + e_r \quad ...(10.7.3)$$

Substituting the values of currents obtained from Eq. (10.7.1) in Eq. (10.7.2) we get

$$\frac{e_t}{Z_2} = \frac{e_f}{Z_1} - \frac{e_r}{Z_1} \quad ...(10.7.4)$$

Substituting for e_r from Eq. (10.7.3) in Eq. (10.7.4) we obtain

$$\frac{e_t}{Z_2} = \frac{e_f}{Z_1} - \frac{e_t - e_f}{Z_1}$$

$$\frac{e_t}{Z_2} + \frac{e_f}{Z_1} = \frac{2e_f}{Z_1}$$

$$\left(\frac{Z_1 + Z_2}{Z_1 Z_2}\right) e_t = \frac{2e_f}{Z_1}$$

Therefore, the voltage transmitted in B is

$$e_t = \frac{2Z_2}{Z_1 + Z_2} e_f = \tau e_f \quad ...(10.7.5)$$

where τ = transmission factor for voltage.

$$\tau_v \triangleq \frac{e_t}{e_f} = \frac{2Z_2}{Z_1 + Z_2} \quad ...(10.7.6)$$

The current transmitted in line B is

$$i_t = \frac{e_t}{Z_2} = \frac{2e_f}{Z_1 + Z_2} = \frac{2Z_1}{Z_1 + Z_2} i_f = \tau_i i_f$$

where Z_i = transmission factor for current.

$$\tau_i \triangleq \frac{i_t}{i_f} = \frac{2Z_1}{Z_1 + Z_2} \quad ...(10.7.7)$$

Voltage reflected in line A is given by

$$e_r = e_t - e_f = \frac{2Z_2}{Z_1 + Z_2} e_f - e_f$$

$$e_r = \frac{Z_2 - Z_1}{Z_1 + Z_2} e_f = \rho_v e_f \qquad \text{...(10.7.8)}$$

where ρ_v = reflection factor for voltage

$$\rho_v \triangleq \frac{e_r}{e_f} = \frac{Z_2 - Z_1}{Z_1 + Z_2} \qquad \text{...(10.7.9)}$$

Current reflected in line A is given by

$$i_r = i_t - i_f = \frac{Z_1 - Z_2}{Z_1 + Z_2} i_f = \rho_i i_f$$

where ρ_i = reflection factor for current

$$\rho_i \triangleq \frac{i_r}{i_f} = \frac{Z_1 - Z_2}{Z_1 + Z_2} \qquad \text{...(10.7.10)}$$

It is seen that

$$\rho_i = -\rho_v$$

Example 10.1 An overhead line is connected in series with a cable. The overhead line has an inductance of 2 mH/km and capacitance of 0.01 μF/km. The cable has an inductance of 0.25 mH/km and capacitance of 0.102 μF/km. If a surge having a maximum value of 100kV travels along the overhead line towards its junction with the cable, calculate:

(a) the surge impedances of the line and the cable.

(b) the velocities of wave propagation in the line and the cable.

(c) the reflected and transmitted waves of voltage and current at the junction.

If the 100 kV surge originates in the cable. Calculate the reflected and transmitted waves of voltage and current at the junction.

Solution : Surge impedance of overhead line,

$$Z_1 = \sqrt{\frac{L_1}{C_1}} = \left(\frac{2 \times 10^3}{0.01 \times 10^{-6}}\right)^{1/2} = 447\ \Omega$$

Surge impedance of cable,

$$Z_2 = \sqrt{\frac{L_2}{C_2}} = \left(\frac{0.25 \times 10^{-3}}{0.102 \times 10^{-6}}\right)^{1/2} = 49.5\ \Omega$$

Velocity of the wave in the overhead line

$$u_1 = \frac{1}{\sqrt{L_1C_1}} = \frac{1}{(2 \times 10^3 \times 0.01 \times 10^{-6})^{1/2}} = 2.24 \times 10^5 \text{ km/s}$$

Velocity of the wave to cable,

$$u_2 = \frac{1}{\sqrt{L_2C_2}} = \frac{1}{(0.25 \times 10^{-3} \times 0.102 \times 10^{-6})^{1/2}}$$

$$= 1.98 \times 10^5 \text{ km/s}$$

Waves Originating in Overhead Line

When the waves originate in the overhead line and travel towards the cable we shall take $Z_1 = 447\ \Omega$, $Z_2 = 49.5\ \Omega$

Incident values

Incident voltage in the overhead line, $e_f = 100$ kV

Incident current in the overhead line,

$$i_f = \frac{e_f}{Z_t} = \frac{100 \times 1000}{447} = 223.7 \text{ A}$$

Transmitted values

Line-to-cable transmission factor for voltage,

$$\tau_{v1} = \frac{2Z_2}{Z_1 + Z_2} = \frac{2 \times 49.5}{447 + 49.5} = 0.1994$$

Transmitted voltage in the cable,

$$e_t = \tau_{v1}\, e_f = 0.1994 \times 100 = 19.94 \text{ kV}$$

Line-to-cable transmission factor for current,

$$\tau_{i1} = \frac{i_t}{i_f} = \frac{2Z_1}{Z_1 + Z_2} = \frac{2 \times 447}{447 + 49.5} = 1.8006$$

Transmitted current in the cable, it

$$i_t = \tau_{i1}\, i_f = 1.8006 \times 223.7 = 402.8 \text{ A}$$

Alternatively, $$i_t = \frac{e_t}{Z_2} = \frac{19.94 \times 1000}{49.5} = 402.8 \text{ A}$$

Reflected values

Reflection factor for voltage,

$$\rho_{v1} = \frac{Z_2 - Z_1}{Z_1 + Z_2} = \frac{49.5 - 447}{447 + 49.5} = -0.8006$$

Voltage reflected for the overhead line,

$$e_r = \rho_{v1}\, e_f = -8.006 \times 100 = -80.06 \text{ kV}$$

Reflection factor for current,

$$\rho_{i1} = \frac{Z_1 - Z_2}{Z_1 + Z_2} = -\rho_{v_1} = 0.8006$$

Current reflected back in overhead line,

$$i_r = \rho_{i1}\, e_f = 8.006 \times 223.7 = 179.1 \text{ A}$$

Alternatively, $$i_r = -\frac{e_r}{Z_1} = \frac{80.06 \times 1000}{447} = 179.1 \text{ A}$$

Check

$$e_f + e_r = e_t$$

$$e_t = 100 + (-80.06) = 19.94 \text{ kV}$$

$$i_r = i_t - i_f = 402.8 - 223.7 = 179.1 \text{ A}$$

The voltage and current distributions of the wave reaching the junction of the line and cable are shown in Fig. 10.2

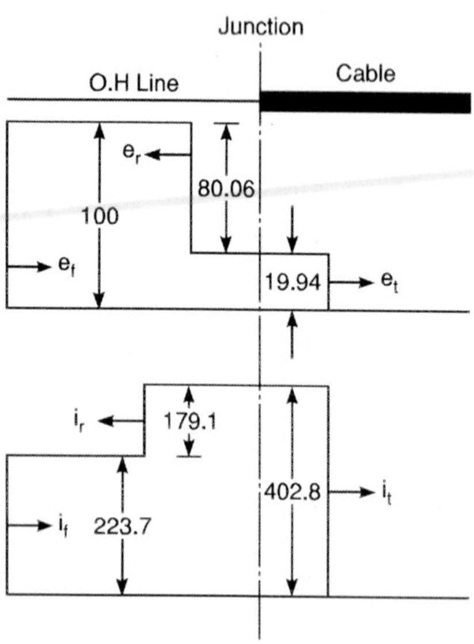

Fig. 10.2 Voltage and current distribution

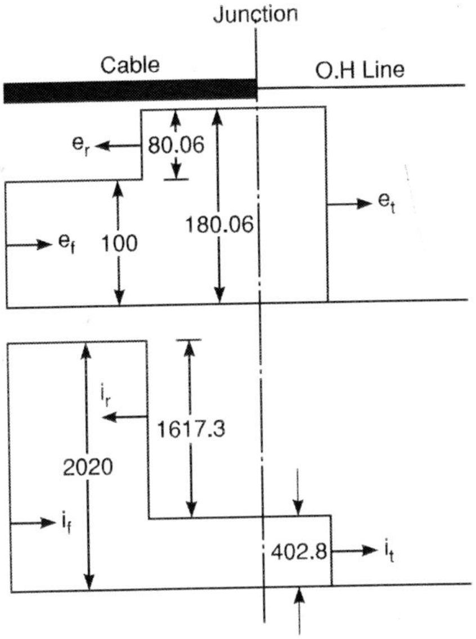

Fig. 10.3 Voltage and current distribution

Waves Originating in Cable

When the waves originate in the cable we shall take

$$Z_1 = 49.5\ \Omega, Z_2 = 447\ \Omega$$

Incident values

Incident voltage in the cable, $e_f = 100$ kV

Incident current in the cable,

$$i_f = \frac{e_f}{Z_1} = \frac{100 \times 1000}{49.5} = 2020 \text{ A}$$

Transmitted values

Cable-to-line transmission factor for voltage,

$$\tau_{v2} = \frac{2Z_2}{Z_1 + Z_2} = \frac{2 \times 447}{49.5 + 447} = 1.8006$$

The voltage transmitted into the overhead line after its arrival at the junction

$$e_t = \tau_{v2}\, e_f = 1.8006 \times 100 = 180.06 \text{ kV}$$

Cable-to-line transmission factor for current,

$$\tau_{i2} = \frac{2Z_1}{Z_1 + Z_2} = \frac{2 \times 49.5}{49.5 + 447} = 0.1994$$

Current transmitted into the overhead line,

$$i_t = \tau_{i2}\, i_f = 0.1994 \times 2020 = 402.8 \text{ A}$$

Alternatively,

$$i_t = \frac{e_t}{Z_2} = \frac{180.06 \times 1000}{447} = 402.8 \text{ A}$$

Reflected values

Reflection factor for voltage,

$$\rho_{v_2} = \frac{Z_2 - Z_1}{Z_2 + Z_1} = \frac{447 - 49.5}{447 + 49.5} = 0.8006$$

Reflected voltage into the cable,

$$e_r = \rho_{v2}\, e_f = 0.8006 \times 100 = 80.06 \text{ kV}$$

Reflection factor for current,

$$\rho_{i_2} = \frac{Z_1 - Z_2}{Z_1 + Z_2} = -\rho_{v_2} = -0.8006$$

Reflected current in the cable.

$$i_r = \rho_{i_2}\, i_f = 0.8006 \times 2020 = -1617.2 \text{ A}$$

Alternatively,

$$i_r = \frac{e_r}{Z_t} = \frac{-80.06 \times 1000}{49.5} = -1617.3 \text{ A}$$

Check

$$e_f + e_r = e_t; \quad 100 + 80.06 = 180.06$$

$$i_f + i_r = i_t; \quad 2020 - 1617.2 = 402.8$$

The distribution of voltage and current waves reaching the junction are given in Fig. 10.3.

10.8 EFFECT OF CABLE ON SURGE

In example 10.1 we have seen that in the first case the surge originates in the overhead line of higher surge impedance and enters into the cable of lower surge impedance. It is found that the transmitted voltage in the cable is much reduced and the transmitted current is increased. In the second case the surge travels from the cable to the overhead line; the transmitted voltage is increased and the transmitted current reduced. The connection of a short length of a cable at the termination of an overhead line with a substation serves to protect the terminal equipment there from dangerous values of voltage rises. The length of cable in the circuit is also important. Unless the cable length is more than the length of the incident wave, the peak value of the wave at the junction will be higher than that of the incident wave. This is due to the repeated reflections at the junction. Thus, instead of absorbing the surge, the cable of short length shall prove more dangerous in the matter of building up of voltage which is greater than the original incident voltage wave. The cable reduces the front steepness of a surge, and reduces it as peak value depending on the ratio of the propagation time in the cable to the duration of surge.

Example 10.2 An overhead line of surge impedance 400 ohms is connected in series with a cable of surge impedance 40 ohms. The cable is 5 km in length and is open at its far end. A surge of peak value 100 kV travels along the overhead line towards the junction of the line and the cable. Find the voltage distribution at the instants 1 and 50 microseconds after the surge reaches the junction. The velocities of propagation in the overhead line and the cable are 3×10^5 km/s and 1.4×10^5 km/s respectively.

Solution In the problem

$$Z_1 = 400\ \Omega,\ Z_2 = 40\ \Omega$$

Line-to-cable reflection coefficient,

$$\rho_1 = \frac{Z_2 - Z_1}{Z_2 + Z_1} = \frac{40 - 400}{40 + 400} = -0.818$$

Reflected voltage in overhead line,

$$e_r = \rho_1\, e_f = -0.818 \times 100 = -81.8 \text{ kV}$$

Transmission factor for voltage

$$\tau_v = \frac{2Z_2}{Z_1 + Z_2} = \frac{2 \times 40}{400 + 40} = 0.812$$

Therefore the voltage transmitted into the cable,

$$e_t = \tau_v\, e_f = 0.182 \times 100 = 18.2 \text{ kV}$$

Alternatively, $e_t = e_f + e_r = 100 + (-81.8) = 18.2$ kV

The voltage at the junction just after the surge strikes the junction is, therefore, 18.2 kV.

The distance travelled by the transmitted wave into the cable is one microsecond

$$= 1.4 \times 10^8 \times 10^{-6} = 140 \text{ m}$$

The distance travelled by the reflected wave into the overhead line in one microsecond

$$= 3 \times 10^8 \times 10^{-6} = 300 \text{ m}$$

The voltage distribution after 1μs is given Fig. 10.4.

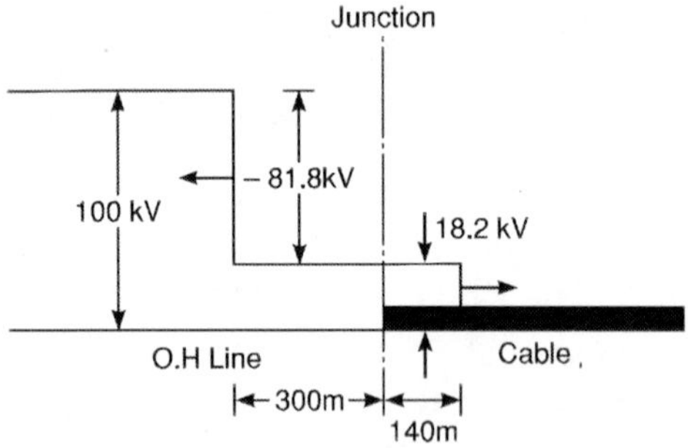

Fig. 10.4 Voltage distribution after 1μs

Time required by the transmitted wave to reach the open end of the cable

$$= \frac{5 \times 1000}{1.4 \times 10^5 \times 10^3} \text{ s} = \frac{5 \times 10^6}{1.4 \times 10^5} = 35.7 \text{ μs}$$

When the voltage reaches the open end of the cable it is reflected and doubled. In the remaining time of (50 – 35.7), i.e, 14.3 μs the distance covered by the wave reflected at the open end is $(14.3 \times 10^{-6} \times 1.4 \times 10^5)$ i.e., 2 km from the free end of the cable towards the junction.

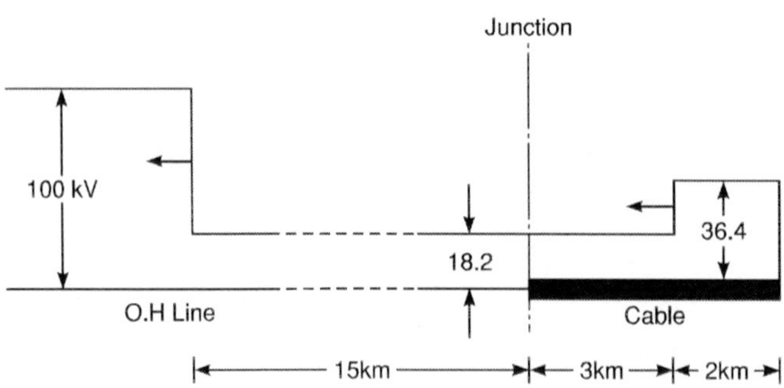

Fig. 10.5 Voltage distribution after 50μs

The distance covered by the reflected wave in the overhead line is 50 μs

$$= 3 \times 10^5 \times 50 \times 10^{-6} = 15 \text{ km}$$

The voltage distribution after 50 μs the original surge reaches the junction of the line and the cable is shown in Fig. 10.5

10.9 JUNCTION OF SEVERAL LINES

Now consider the case of a surge travelling along a line of surge impedance Z_1 reaching a junction of several lines of surge impedance Z_2, Z_3, Z_4 etc. The wave will be partially reflected on the line of surge impedance Z_1 and partially transmitted in other branches. For the sake of simplicity, let a line *OA* be bifurcated into two branches *AB* and *AC* (Fig. 10.6) of surge impedances Z_2 and Z_3 respectively. Since *AB* and *AC* are in parallel, the voltages transmitted in them will be the same.

Let i_1 and i_2 be the currents transmitted in branches *AB* and *AC*. The following relations may be written on the basis of the arguments in the previous sections:

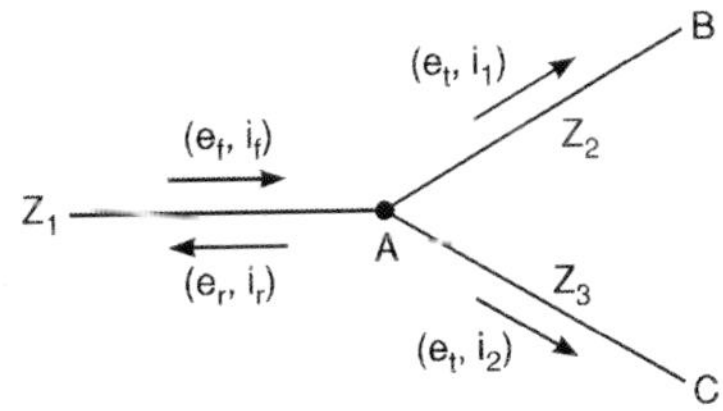

Fig. 10.6 Forked line

$$\left.\begin{aligned} i_f &= \frac{e_f}{Z_1} \\ i_r &= \frac{-e_r}{Z_1} \\ i_1 &= \frac{e_t}{Z_2} \\ i_2 &= \frac{e_t}{Z_3} \end{aligned}\right\} \quad \ldots(10.9.1)$$

The voltage at the junction is

$$e_f + e_r = e_t \qquad ...(10.9.2)$$

Also

$$i_f + i_r = i_1 + i_2 \qquad ...(10.9.3)$$

Substituting the values from Eq. (10.9.1) in Eq. (10.9.2)

$$\frac{e_f}{Z_1} - \frac{e_r}{Z_1} = \frac{e_t}{Z_2} + \frac{e_t}{Z_3} \qquad ...(10.9.4)$$

Putting the values of e_r from Eq. (10.9.2) in Eq. (10.9.4)

$$\frac{e_f}{Z_1} - \frac{e_t - e_f}{Z_1} = \frac{e_t}{Z_2} + \frac{e_t}{Z_3}$$

which on simplification gives

$$e_t = \frac{\frac{2e_f}{Z_1}}{\frac{1}{Z_1} + \frac{1}{Z_2} + \frac{1}{Z_3}} = 2e_f \frac{\frac{1}{Z_1}}{\frac{1}{Z_1} + \frac{1}{Z_2} + \frac{1}{Z_3}} \qquad ...(10.9.5)$$

The reflected voltage,

$$e_r = e_t - e_f = e_f \frac{\frac{1}{Z_1} - \frac{1}{Z_2} - \frac{1}{Z_3}}{\frac{1}{Z_1} + \frac{1}{Z_2} + \frac{1}{Z_3}} \qquad ...(10.9.6)$$

Example 10.3. A cable with a surge impedances of 100 ohms is terminated in two parallel-connected open-wire lines having surge impedances of 600 and 1000 ohms respectively. If a steep- fronted voltage wave of 1000 V travels along the cable, find from the first principles the voltage and current in the cable and the open-wire lines immediately after the travelling wave has reached the transition point. The line may be assumed to be of infinite length.

Solution : In the problem

$$Z_t = 100\ \Omega,\ Z_2 = 600\ \Omega,\ Z_3 = 1000\ \Omega, \text{ and } e_f = 1000\ \text{V}$$

$$i_f + i_r = i_1 + i_2$$

$$\frac{e_f}{Z_1} - \frac{e_r}{Z_1} = \frac{e_t}{Z_2} + \frac{e_t}{Z_3}$$

$$e_f - e_r = e_t\left(\frac{1}{Z_2} + \frac{1}{Z_3}\right)Z_1$$

$$= e_t\left(\frac{1}{600}+\frac{1}{1000}\right)\times 100 = \frac{4}{15}e_t$$

Also, $\quad e_f + e_r = e_t$

By addition $\quad 2e_f = \frac{4}{15}e_t + e_t = \frac{19}{15}e_t$

$$e_t = \frac{15\times 2}{19}e_f = \frac{30}{19}\times 1000 = 1579 \text{ V}$$

Reflected voltage, $e_r = e_t - e_f = 1579 - 1000 = 579$ V

Current transmitted in the 600 ohm line,

$$i_1 = \frac{e_1}{Z_2} = \frac{1579}{600} = 2.63 \text{ A}$$

Current transmitted in the 1000 ohm lime,

$$i_2 = \frac{e_1}{Z_3} = \frac{1579}{1000} = 1.58 \text{ A}$$

Current in the cable $= i_1 + i_2 = 2.63 + 1.58 = 4.21$ A

Reflected current $i_r = -\frac{e_r}{Z_1} = \frac{-579}{1000} = -5.79$ A

Incident current $i_f = -\frac{e_f}{Z_1} = \frac{1000}{100} = 10$ A

Current in the cable, $i_f + i_r = 10 - 5.79 = 4.21$ A

This result is the same as obtained above. It serves as a check.

10.10 POWER SYSTEM OVERVOLTAGES

Power systems may be subjected to overvoltages which may be of transient or persistent nature. Overvoltages stress the insulation of the equipment beyond safe limits. Their study, therefore, serves as a guide in the choice of insulation level and the requirements for protective devices so as to minimize damage of equipment. The main causes due to which overvoltages are produced in the power systems may be conveniently grouped into two categories, namely, internal and external.

10.11 INTERNAL OVERVOLTAGES

The internal overvoltages have got their origin within the system itself. They are further classified as transient, dynamic and stationary. The transient overvoltages persist for short durations. They may be caused by sudden changes in the system parameters, usually as a result of switching-in or tripping-off of highly inductive or capacitive components of the system such as transformers, shunt reactors, and long transmission lines. The magnitude of the switching surges depends on the system voltage, the design of the circuit breakers, and the circuit constants. For a well designed system their magnitude normally does not exceed three to four times the working voltage. The nature of these transient voltages is often oscillatory and their waveforms vary considerably. They are propagated along the line as travelling waves and reach the terminal equipment. The insulation of the line and terminal equipment is thus subjected to high voltages. Switching surges have relatively greater importance at high voltages.

Overvoltage is also produced due to arcing faults on systems having insulated neutrals. An ***insulated neutral system*** is a system which has no intentional connection to earth except through indicating, measuring or protective devices of very high impedance. On such a system if one of the phase breaks down to earth due to failure of the insulation, the voltage between sound conductors and earth may build up to high values. This is due to the fact that the faulty conductor comes in contact with earth, the voltage between sound conductors and earth becomes the line voltage. An arc is formed between earth and faulty conductors at the point where the insulation failed. With the formation of the arc the voltage across it becomes zero and thereby the arc is extinguished, the potential of the faulty conductor is restored and a formation of second arc takes place. This phenomenon is called ***arcing ground***. The successive extinction and re-ignition of the charging current flowing in the arc may increase the potential of the other two healthy conductors appreciably due to setting up of high frequency oscillations. The overvoltage in healthy conductors may damage the insulation at some other point and thus resulting in a short circuit on the system. For this reason, it is generally avoided to operate high voltage systems with insulated neutrals.

Overvoltages produced at normal system frequency and persisting only for a few seconds are called ***dynamic overvoltages***. Such voltages are encountered in generators of hydro-power plants which overspeed on disconnection or removal of a large part of load from them.

Stationary overvoltages are also, produced at system and persist for long periods, perhaps for hours. These overvoltages are produced on healthy conductors under sustained earth fault conditions on one conductor in the system which is earthed through an arc suppression coil.

10.12 EXTERNAL OVERVOLTAGES

Lightning is very serious cause of overvoltages on the lines. It is advisable at this stage to know the nature of these voltages. Lightning denotes the electrical discharge phenomena in the atmosphere which are essentially associated with thunderstorms. The discharge may take place between the thundercloud and earth, between different thunderclouds, or even between different charges in the same thundercloud. Only the discharges from the clouds to earth are of interest to electrical engineers because such discharges affect the transmission lines and equipment.

10.13 MECHANISM OF THE LIGHTNING DISCHARGE

Many theories have been put forward to account for the accumulation of charge in clouds. The subject is still controversial. But it has been established, irrespective of the process of charge accumulation that during thunderstorms there is a concentration of negative charge in the lower layers of the cloud, while the upper zone becomes positively charged (Fig. 10.7)

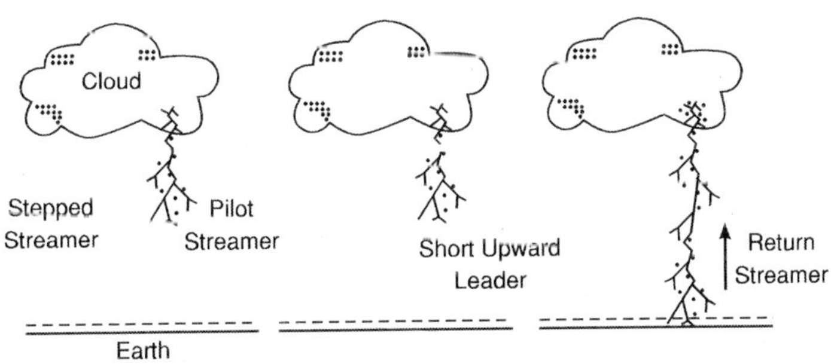

Fig. 10.7 Mechanism of lightning stroke

When the voltage gradient in the cloud is much greater than that at the earth's surface beneath, the discharge originates in the cloud. When the potential gradient, which is not uniform in a cloud, becomes of the order of 10-30 kV per cm in any part of the cloud, an initial discharge, called the ***pilot streamer*** or ***pilot leader***, moves slowly from it towards the earth. The first discharge moves to earth in steps of about 50 metres each and is, therefore, termed the ***stepped leader*** (Fig. 10.7). The pilot leader carries a charge with it, and the potential gradient at its tip is very high. It ionizes the air and provides a path or channel for the pilot leader. The channel also becomes charged. The plot leader carries with it the secondary streamers, which branch out from it. As the pilot streamer approaches the earth, the electric field intensity increases and due to this a charge of opposite polarity in the form of a short streamer rises from the earth to meet the tip of downward leader. When a contact is made between the pilot leader and short upward streamer, a return streamer travels from the earth to cloud along the ionized channel formed by the pilot leader (Fig. 10.7). The return streamer moves very fast and produces the well known intensely luminous lightning flash.

The potential of the portion of the cloud where the discharge originated is lowered by the passage of the charge through the ionized channel to the earth. But other portions of the cloud remain charged. Therefore, a high potential develops between original charge centre and another charge centre in the cloud. The charge of the second charge centre is first transferred to the first one, and then it passes to the earth through the ionized channel made by the first discharge. The second charge is unbranched and without steps. Its velocity is therefore, more than that of the pilot leader. This is known as ***dart leader*** and it is also followed by a return stroke.

Other charges are similarly discharged to the earth in the form of leader and return strokes along the same ionized channel. Lightning strokes with many dischargers are known as ***multiple*** or ***repetitive strokes***. But only a single flash is seen with an unaided eye. As many as forty components strokes have been recorded in such multiple strokes.

The lightning strokes are referred to as ***hot*** or ***cold*** depending upon the magnitude and duration of the stroke current. The stroke of hot lightning has a low current but long duration. It causes fire when it strikes an object. The stroke of cold lightning has a high current but is of short duration. It causes an explosion when it strikes an object.

10.14 FREQUENCY OF LIGHTNING FLASHOVERS

The frequency of lightning flashovers to overhead lines at a place is important to know the probability of outages due to lightning. In a particular region there is a wide variation in the frequency of thunderstorms, but the average frequency is somewhat constant over a period of years. The number of days per year on which thunder is heard is defined as the ***isokeraunic level***. This number gives an idea of the severity of lightning at a particular place. The isokeraunic level does not give any information about duration and number of strokes reaching ground. The isokeraunic level in Great Britain is about 15, and in India it is generally between 20 to 35 with the exception of extreme south (Cape Comorin) and regions of Assam and Bay of Bengal. In tropical zones the values are higher.

10.15 STROKE CURRENTS

The instruments commonly employed for measuring lightning stroke current are klydonograph, magnetic link, surge crest ammeter, laboratory oscillographs and fulchronograph. With the help of these instruments it has been found that about 90 percent of the lightning strokes have a negative polarity, that is, the cloud has negative and earth positive polarity. The magnitude of the lightning current varies from one thousand amperes to about 200 kA or even more; the average value is about 20 kA.

10.16 DIRECT AND INDUCED STROKES

Lightning overvoltages are produced in the lines either by a direct stroke or by an induced stroke. When lightning strikes a line conductor, the current divides and flows in each direction along the line. The line is subjected to a voltage given by the product of the lightning current and half the surge impedance of the line. This voltage causes a flashover of the line insulation to earth. It has been observed that the risk of lightning flashovers goes on diminishing with the increase in operation voltages.

Overvoltages can also be induced on overhead lines by strokes to earth or an earthed object, near the line. Such strokes are known as ***induced strokes***. Induced strokes are also called ***indirect strokes***. The voltages due to induced strokes are very much less as compared to those produced by direct strokes.

10.17 WAVESHAPES OF STROKE CURRENTS

The wave shape of stroke current is also important. It varies widely. For testing electrical equipment arbitrary wave shapes having the nature of uni-directional impulses have been standardized. Such a wave shape can be produced by impulse generator. The wave shape consists of a portion showing the steep rise of voltage upto a peak or crest value called the ***wavefront***, and the other portion showing the decay of voltage called the ***wavetail***. Such a wave is shown in Fig. 10.8. This maybe represented as the difference of two exponentials, thus

$$e = E(\varepsilon^{-\alpha} - \varepsilon^{-\beta})$$

where α and β are constants which determine the shape.

The wave is defined by times t_1 and t_2 in microseconds. The time to reach the impulse current or voltage to 90 percent of its maximum amplitude is denoted by t_1 and t_2 denotes the time when the current or voltage has fallen to half of its peak value. Both the timings are measured from the start of the wave. This is designated as t_1/t_2 or t_1 and t_2. In Europe 1/50 microsecond, while in the USA 1.5/40 microsecond waveshapes have been standardized for testing purpose. The International Electrotechnical Commission recommends 1/50 microsecond wave as standard. In India 1.2/50 microsecond is the standard.

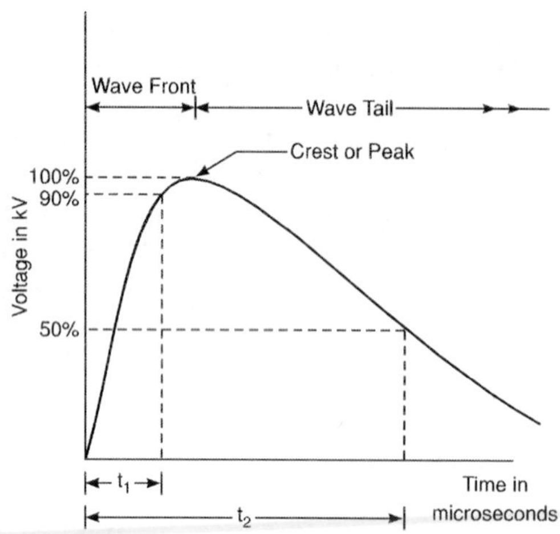

Fig. 10.8 Impulse waveform

10.18 OVERVOLTAGE PROTECTION

During the travel the voltage surges suffer a reduction in magnitude and steepness. This may result in a reduced stress on insulation. Even after attenuation these overvoltages may cause damage to insulators and substation equipment. It is therefore, necessary to provide means to protect the insulators and other apparatus from the harmful effects of the surge voltages. A number of devices are available to reduce the amplitude and front steepness of surges. Of the various devices adopted for overvoltage protection the following will be described here:

1. Overhead earth wire
2. Rod gap
3. Surge diverter

10.19 OVERHEAD EARTH WIRE

An ***overhead earth wire*** or ***ground wire*** is one of the most common devices used to protect the line against the lightning. It is a wire carried by the line supports and runs over the phase conductors (Fig 10.10). The purpose of the earth wire is to intercept direct lightning strokes, which would otherwise strike the phase conductors. It has no effect on switching surges. When lightning strikes an earth wire at mid span, waves are produced which travel in opposite directions along the line. The waves reach the adjoining towers which pass them to earth safely. It should be noted that the earth wire is effective only when the resistance between the tower footing and earth is sufficiently low. In case the resistance is not low, the earth wire or tower struck by lightning will be raised to a very high potential, which will cause a flashover from the tower to one or more phase conductors. Such a flashover is known as ***back flashover***. It occurs only when the product of tower current and tower impedance exceeds the insulation level of the line. Back flashovers on the line can be minimized by reducing tower footing resistance using driven rods and counterpoises where soil resistivity is high. A ***counterpoise*** is a conductor buried in the ground. Counterpoise wires are usually made

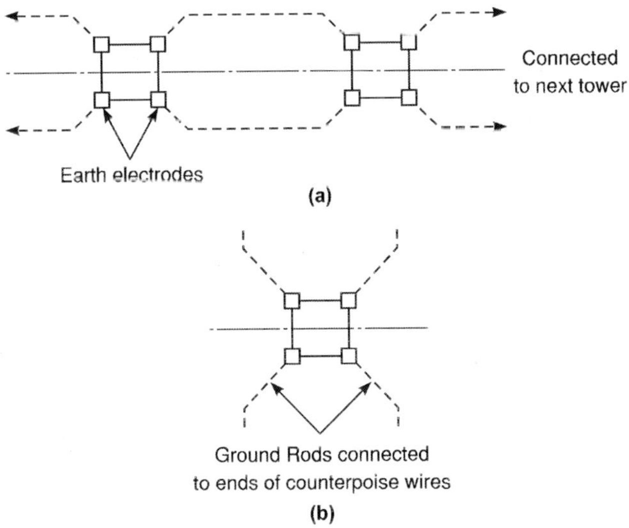

Fig. 10.9 Types of counterpoises (a) Parallel; (b) Radial

of galvanized steel. The counterpoise for an overhead transmission line consists of a special grounding terminal that reduces the surge impedance of the ground connection and increases the coupling between the ground wire and the conductors. Two types of counterpoises namely, parallel type and ring type, are used for transmission lines. The ***parallel*** (or ***continuous***) ***counterpoise*** is made up of one or more counterpoise wires buried under the transmission line throughout its length. The counterpoise wires are connected to the overhead earth wire at all towers or poles. The ***radial type counterpoise*** is made up of number of wires extending radially from the tower legs. The number and length of the wires are determined by the tower location and soil conditions. Fig. 10.9 shows the two types of counterpoises. The depth of counterpoises may be 0.5 m to 1.0 m below the surface of the earth.

The tower footing resistance of 30 Ω for 132 kV lines and less than 20 Ω for 275 and 400 kV lines to avoid risk of back flashovers have generally proved satisfactory in most cases.

The efficacy of an earth wire to shield a conductor depends on its shielding angle. The ***shielding*** or ***protective angle*** α is the angle between the vertical through the earth wire and phase conductor to be protected (Fig. 10.10). Usually the angle included between the vertical through the earth wire and the line joining the earth wire to the outer most phase conductor is taken as the shielding angle. For effective shielding the protective angle should be kept as small as possible. Adequate shielding is achieved provided that this angle does not exceed 40° . The angles between 20°–30° are quite safe.

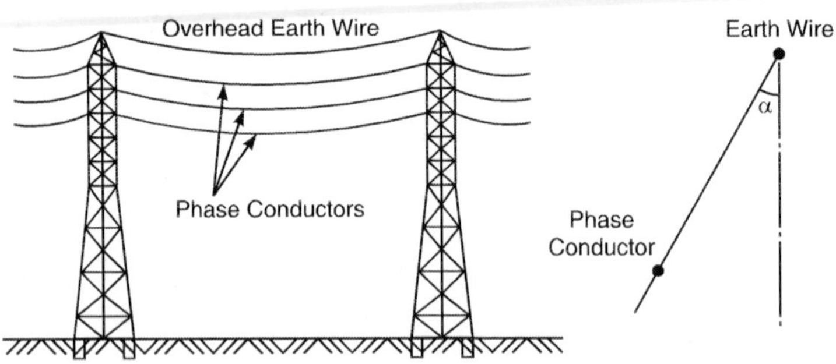

Fig. 10.10 Overhead earth wire and shielding angle

Two earth wires are used in modern high voltage systems with wider spacing between the conductors. The protection afforded by two earth wires is definitely better than a single earth wire. Moreover, the surge impedance is low for two earth wires and the coupling effect with the conductor is increased.

The earth wire arrangement is useful in front of high voltage stations for one or two kilometers on the line for extra protection. Both electrical and mechanical considerations are made to select an earth wire. It should have a current capacity to carry heavy short circuit currents or the currents when the phase conductors are earthed. It should also be sufficiently strong mechanically. Steel cored aluminium conductors or alumoweld conductors may be used as earth wires.

10.20 ROD GAP

The rod gap is one of the most common forms of protective devices. It is an air gap between ends of two rods. The rod gaps are generally used on transformer bushings and at the ends of strings of suspension insulators. They are shown in Figs. 10.11 and 10.12. The gap setting should be such that it should break under all conditions before the equipment to be protected is affected. The chief merits of this device are simplicity, reliability and cheapness. Although rod gaps provide a certain degree of protection against the voltages they got the following limitations:

(a) They are unable to prevent the flow of power follow current which flows in the gap after the breakdown. The flow of this current sets up an arc across the gap which is extinguished by

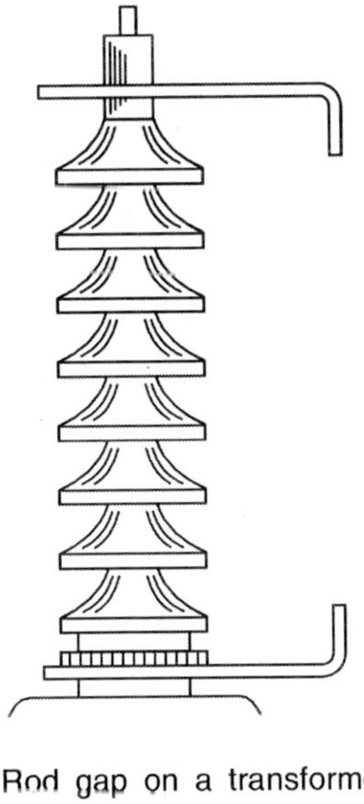

Fig. 10.11 Rod gap on a transformer bushing

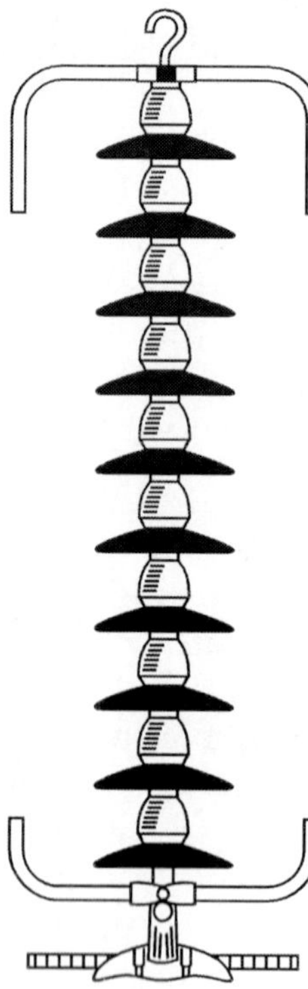

Fig. 10.12 Rod gap on an insulator bushing

the operation of the protective gear resulting in the supply interruption.

(b) The rod gap has a bent sparkover voltage-time characteristic. If it is set to protect the equipment at the long times it would not do so for short times. If it is set for protecting the equipment at the short times, its minimum spark-over voltage will be too low and cause frequent flashovers. The supply will, therefore, be interrupted even due to harmless surges. A compromise is, therefore, made in gap setting.

(c) Atmospheric conditions such as humidity, temperature and barometric pressure affect the performance of gaps.

(d) The arc set up may produce sufficient heat to damage the rod gap by melting it and thus alter the gap setting.

(e) The polarity of the surge also affects the performance of the gap.

The rod gaps may be used in practice where the continuity of supply is not of much importance. In such cases they are used with automatic circuit breakers, which will operate as soon as the arc current has been interrupted. Despite the drawbacks rod gaps are used with surge diverters so as to protect the equipment to some degree in case of failure of surge diverters.

10.21 SURGE DIVERTERS

Surge diverters or lightning arrester is a device used for diverting abnormal high voltages to grounds without affecting the continuity of supply. The name surge diverter appears to be more correct than lightning arrester. Three types of surge diverters are described here:

1. Expulsion type surge diverter
2. Valve-type surge diverter
3. Metal-oxide surge diverter

10.22 EXPULSION TYPE SURGE DIVERTER

The expulsion type diverter Fig.10.13 consists essentially of a tube made of fibre or other gas producing material. The metal electrodes are provided at the ends of the tube forming a spark gap inside the tube. The electrode at the upper end is separated from the line conductor to form an external series gap with the conductor. Under the normal working conditions this gap avoids the leakage currents through the tube. The bottom electrode, which is hollow, is connected to earth.

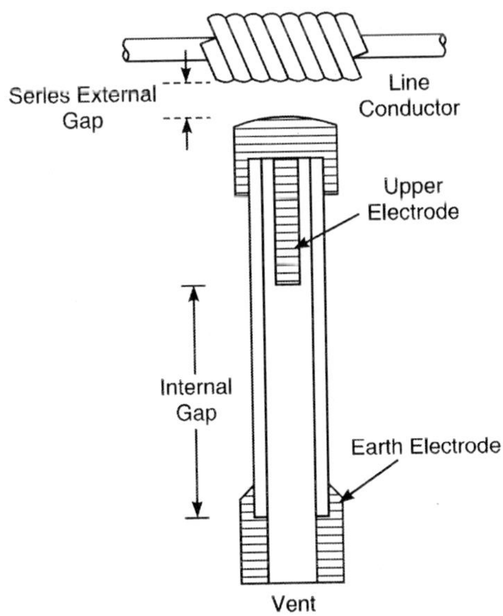

Fig. 10.13 Expulsion type surge diverter

When a high voltage appears on the conductor both external and internal gaps spark-over. A gas is produced by the vaporization of the fibre tube walls. The great pressure developed in the tube flows out the arc through the vent formed by the bottom electrode. The expulsion of the arc takes place so violently that the arc is interrupted when the current passes through zero point. If the rate of recovery of dielectric strength of the gap is more than the rate of rise restriking voltage, the restriking of the arc is prevented, and thus there is no interruption in supply.

Although, it has been possible to design the expulsion gap to extinguish the arc at the smallest earth fault currents in high voltage systems, there is an upper limit for the current to be interrupted. If that limit is exceeded it may generate excessive gas which may produce sufficient pressure to cause mechanical failure of the tube, or the insulating material may be eroded to a considerable amount.

The expulsion tube is an improvement over the rod gap. The impulse sparkover characteristic is better than that of the rod gap. The expulsion tubes are mostly used on transmission lines. They may be used for protecting distribution type of transformers where economy and continuity of supply are the main considerations. The use is mainly restricted to system with a low fault level.

10.23 VALVE-TYPE SURGE DIVERTER

Valve-type surge diverter is also known as conventional gap surge diverter or silicon carbide (SiC) surge diverter with series gaps. This type of diverter is commonly known by trade names such as thyrite, miurite, magnavalve, etc.

The valve-type surge diverter (Fig. 10.13) consists of a multiple spark gap assembly in series with resistor elements of nonlinear volt-ampere (V-I) characteristic. Each spark gap has two electrodes. In order to have uniform voltage distribution, between the gaps, nonlinear resistors are connected in parallel across each gap. Resistor elements are made of silicon carbide (SiC) with inorganic binders. The material may be in the form of pellets, discs, or blocks. The whole arrangement is enclosed in a sealed porcelain or pyrex glass housing filled either nitrogen gas or SF_6 gas.

Silicon carbide resistor units have volt-ampere (V–I) characteristic given by the relationship

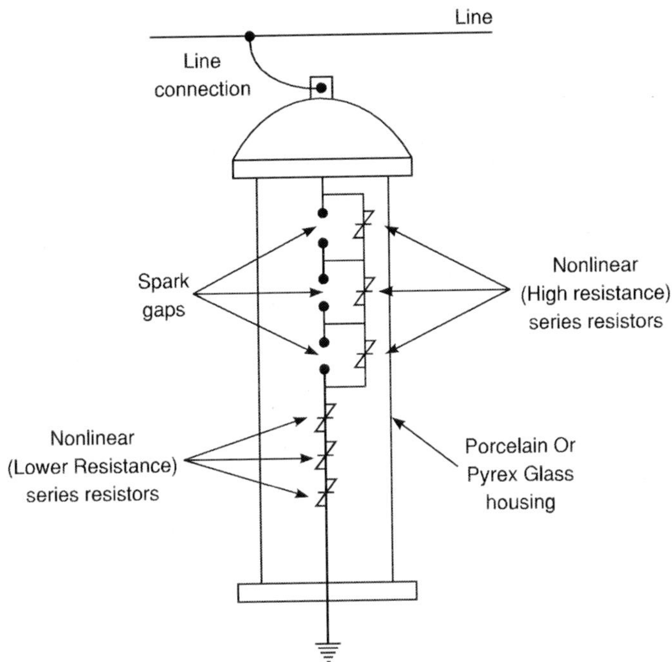

Fig. 10.14 Valve type surge diverter

$$I = kV^n$$

where n varies between 5 and 8.

When a voltage surge high enough to endanger the equipment insulation appears across the terminals of the diverter, the air gaps spark over, and the current is discharged to ground through the nonlinear resistor which offers very small resistance. After the passage of the surge, the impressed voltage across the diverter falls and the diverter resistance increases until normal voltage is restored. When the surge disappears a small current at power frequency flows in the path produced by the flashover. This current is known as ***power-follow current***. The magnitude of this current is reduced to a value which can be interrupted by the spark gaps as they recover their dielectric strength. The power-follow current is extinguished at the first current zero. The supply therefore, remains uninterrupted. The normal condition of the diverter is restored and it is then ready for further operation with incoming surges. This is called resealing of the surge diverter.

Valve type surge diverters were used earlier in ac systems. At present their use is not very common.

10.24 METAL-OXIDE SURGE DIVERTERS

With the development of metal-oxide resistors, the need for series spark gaps in diverters has disappeared. At present, gapless surge diverters are very widely used. Modern Zinc Oxide (ZnO) diverters provide protection against all types of ac and dc overvoltages. They are used for overvoltage protection at all voltage levels in a power system.

Metal oxide surge diverters are also known as ***gapless surge diverters***, or ***zinc oxide*** (ZnO) ***diverters***. A ZnO diverter is shown in Fig. 10.15. The base material used in manufacturing metal oxide resistors is zinc oxide (ZnO). It is a semicoducting material of N-type. It is pulverized and finely grained. About 10 or more doping materials are added in the form of fine powders of insulating oxides such as Bi_2O_3, Sb_2O_3, CoO, MnO_2, Cr_2O_3. The actual composition differs from one manufacturer to the other. After treating the powder with a number of processes, the mixture is spray dried to obtain dray granulates. The granulates are compressed into disc shaped blocks. The blocks are sintered to obtain a dense polycrystalline ceramic. Each disc metal oxide resistor is coated with a conducting compound in order to protect the discs from undesirable environmental effects. The conducting coating also provides

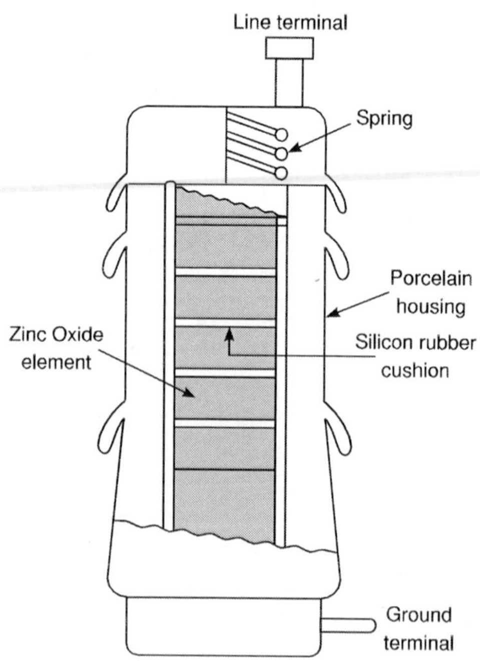

Fig. 10.15 ZnO surge diverter

proper contact and uniform current distribution. The discs are then enclosed in porcelain housing filled with nitrogen gas or SF_6 gas. Silicon rubber is used to keep the discs in positions. It also helps in heat transfer from discs to the porcelain housing. The discs are held under pressure using suitable springs.

The ZnO diverter material has extremely nonlinear voltage-current characteristic given by

$$I = kV^n$$

where n may vary between 35 and 50. The V–I characteristic of a typical ZnO disc is shown in Fig. 10.16.

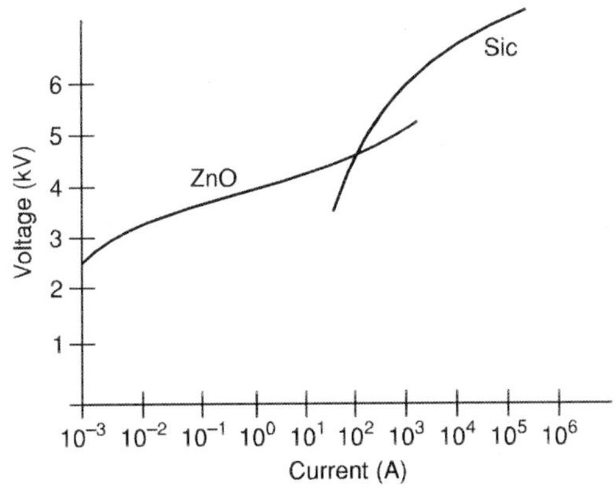

Fig 10.16 Typical V–I characteristic of a ZnO element

The leakage current in a ZnO element at normal voltage is of the order of 100 μA. The same element can carry thousands of amperes at twice the normal operating voltage. This property makes it possible to eliminate series spark gaps in the diverter. It is also seen from the V–I curve that above certain voltages, the characteristic is almost flat. Therefore, the characteristic of the ZnO diverter can be precisely selected with reference to the insulation withstand level characteristic of the protected equipment.

The voltage drop in the ZnO diverter takes place at the grain boundaries. There exists a potential barrier at the boundary of each grain of ZnO. This potential barrier controls the flow of current from one grain to the next. At normal voltages, the potential barrier does not allow the current to flow. When an overvoltage occurs, the barrier collapses and sharp transition from insulating to conducting state takes place. The current starts flowing and the surge is diverted to ground. After the passage of the surge, the impressed voltage across the diverter falls, the current is reduced to a negligible value by the resistor units and there is no power-follow current.

Modern ZnO diverters provide protection against all types of ac and dc overvoltages. They are used for overvoltages protection at all voltage levels in a power system.

10.24.1 Advantages of ZnO Diverters

The ZnO diverters have the following merits.

1. The construction of ZnO surge diverters is simpler than that of SiC diverters with series gaps. The number of parts in ZnO diverter is about 35% as compared to SiC diverter. The weight of ZnO diverter is about 45% of the weight of SiC diverter. The height of ZnO diverter isles than that of SiC diverter.
2. The absence of spark gaps in ZnO diverter eliminates sparkovers and the risk of shock to the system when the gap breaks down. It also eliminates the need of voltage grading system
3. During normal voltage operating conditions the leakage current in ZnO element is very low of the order of 100 μA while it is of the order of 100 A in SiC valve resistor.
4. There is no power-follow current in ZnO diverter.
5. ZnO diverters have higher energy absorbing capability.
6. ZnO diverters possess high stability during and after prolonged discharges.
7. It is possible to control dynamic overvoltages in addition to switching surges. This results in economic insulation coordination.

These advantages have made it possible to use modern ZnO surge diverters for protection against all types of ac and dc overvoltages.

10.25 LOCATION OF SURGE DIVERTERS

Surge diverters are installed close to the equipment to be protected. They are usually connected between phase and ground in ac systems and between pole and ground in dc systems. In ac systems, diverters are provided for each phase. In EHVAC systems, surge diverters are used to protect generators, transformers, busbars, line, circuit breakers, etc. In HVDC systems, the diverters used to protect buses, valves, converter units, reactors, filters, etc.

10.26 SELECTION OF SURGE DIVERTERS

The important factors to be considered in the selection of surge dividers are as follows:

1. Continuous operating voltage (COV)
2. Temporary overvoltage (TOV)
3. Rated voltage
4. Switching surge discharge voltage
5. Energy dissipation capability.

10.27 LIGHTNING PERFORMANCE OF LINES

The lightning performance of EHV lines is dependent on many factors. The basic considerations made in the design of transmission line based on direct lightning strokes are as follows:

1. The lines must be provided with earth wires. The shielding angle should lie between 20° and 30°.
2. Surge diverters should be located properly.
3. Adequate clearance should be provided between phase conductors and tower (ground).
4. There should be sufficient clearance between phase conductors and earth wire specially at midspan. This would eliminate flashovers between conductors and earth wire during conductor swings, dancing, galloping, etc.
5. The tower footing resistance should be as low as economically justified.

10.28 BASIC IMPULSE INSULATION LEVEL OR BASIC INSULATION LEVEL (BIL)

Basic impulse insulation level (BIL) can be defined as reference level expressed in impulse crest (peak) voltage of a standard wave not longer than 1.2/50 microsecond wave, according to Indian standards.

Apparatus insulation levels as demonstrated by suitable tests should be equal to or greater than the BIL.

The basic insulation level measures the ability to withstand test voltages without disruptive discharge. The BIL for a system is selected such that the system should be protected with a suitable lightning protective device, for example, a surge diverter. The margin between the BIL and the surge diverter should be selected such that it provides good protection economically. The BIL of the equipment insulation chosen must be higher than the maximum expected surge voltage across the surge diverter (selected for protection).

Critical Flashover (CFO) Voltage

Critical flashover (CFO) voltage is the peak voltage for a 50% probability of flashover or disruptive discharge.

Impulse Ratio

Impulse ratio (for flashover or puncture of insulation) is the impulse peak voltage divided by the peak value of the power frequency voltage to cause flashover or puncture.

Each equipment is normally tested for both the normal ac frequency and impulse strength.

10.29 VOLT-TIME CHRACTERISTIC (CURVE)

The volt-time curve is graph of the crest flashover voltages plotted against time to flashover for a series of impulse applications of a given waveshape.

For plotting volt-time curve of a given insulation, a series of application of impulse voltages of the same waveshape and polarity but different peak values is carried out. Suitable test waveshape and polarity is applied through an impulse voltage generator. This impulse voltage peak is increased in constant steps of voltage. The largest value of peak voltage at which the given insulation does not flashover is called ***full wave impulse withstand voltage***. If the peak voltage is further increased such that the insulation under test fails to flashover on percent of applications and flashes over the tail of the wave on the other 50 percent of the applications, the crest value of this voltage is known as ***critical flashover voltage***. If the insulation under test flashes over exactly at crest value of the test wave, then it is known as ***crest flashover voltage***. The test voltage at which the test insulation flashes over the front of the wave is called ***front flashover voltage***. The construction of the volt-time curve and the terminology associated with impulse testing is shown in Fig. 10.17.

The values of voltages and the time required for flashovers if plotted on V–t axes give volt-time curves for the test insulation.

Rated withstand voltage is always chosen smaller than the critical withstand voltage depending upon the factor of safety required and economy. Insulation coordination helps in adjusting the insulation levels of different items in a power system.

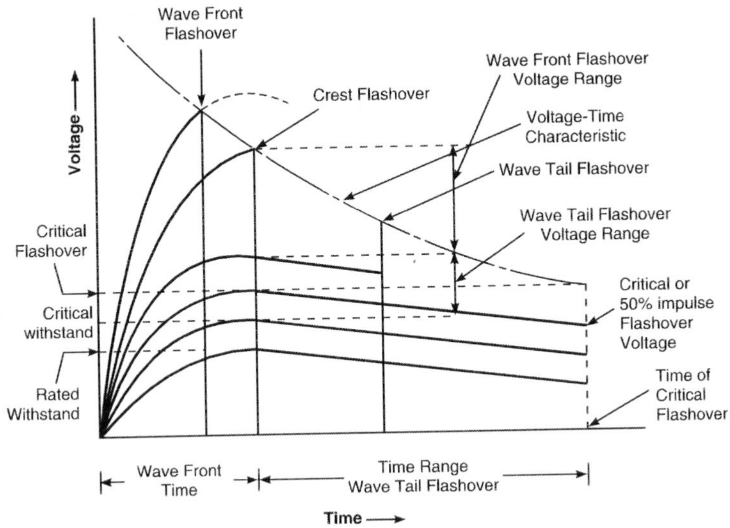

Fig. 10.17 Construction of volt-time curve.

10.30 INSULATION COORDINATION

The equipment used in a power system comprises items having different breakdown or withstand voltages and different volt-time characteristics. Insulation coordination aims at correlating the insulations of these various items with the characteristics of protective devices that the equipment is protected from excessive voltages.

Insulation coordination involves the following basic steps:

1. The determination of the line insulation level.
2. Selection of standard insulation level (BIL) and insulation level of the equipment.
3. Making sure that every equipment has a breakdown strength equal to or higher than the basic impulse level.
4. The selection of proper protective devices that will give the apparatus as good protection as can be justified economically.

Thus, the insulation coordination is the matching of the volt-time flashover and breakdown characteristics of the equipment and protective devices. Proper insulation should ensure that the volt-time curve of the weakest equipment in the system will lie above the volt-time curve of the protective device such as surge diverter, over the whole range of the volt-time curve.

Figure 10.18 shows volt-time curves of various items of the system. The volt-time curve of the substation bus is highest located such that it can even withstand direct lightning strokes.

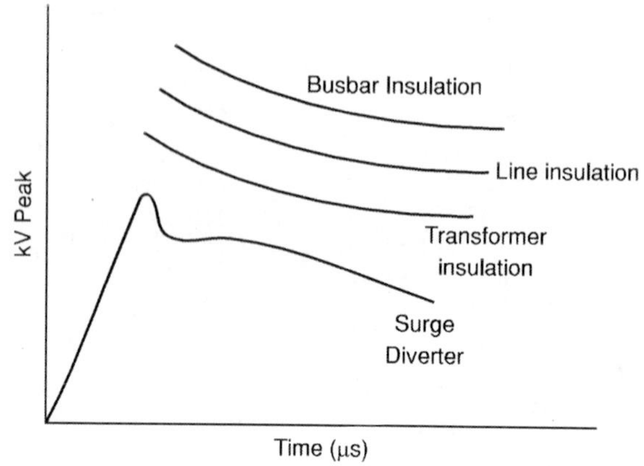

Fig 10.18 Volt-time curves of various items of system

Magnitudes and waveshapes of overvoltages occurring on a power system vary over a wide range. This variation is totally a random phenomenon. Designing of system insulation that could withstand the voltages upto the highest level and providing 100 percent protection is very uneconomical. Insulation failure may occur even with 100 percent protection. For this reason, practical insulation designs use probabilistic and statistical methods considering both economy and safety margins specially for high voltage insulation requirements.

EXERCISES

1. What are the causes of overvoltages in a power system ?
2. Explain the mechanism of the lightning discharge. What are direct and indirect strokes ? Draw and explain the waveshape of stroke currents.
3. What are various devices used for overvoltage protection ?
4. Explain the use of an overhead earth wire for protection of the line against lightning. What is shielding angle ? Explain its significance.
5. Briefly explain the working of rod gap.
6. What are the various types of surge diverters ?

7. Describe the construction and working of expulsion type surge diverter.

8. Describe the construction and working of valve type surge diverter.

9. Describe the construction and working of metal oxide surge diverter.

10. What are the various types of surge diverters ? Explain the working of the gapless surge diverter.

11. Why is insulation coordination needed in a power system ? What is meant by basic insulation level ?

❑❑❑

11

CIRCUIT BREAKING

11.1 INTRODUCTION

The function of a circuit breaker is to make or break an electrical circuit in normal or abnormal conditions. In case of a fault or any abnormal condition, the protective relay sends a tripping signal to the circuit breaker and the circuit breaker isolates the faulty part of the system. The basic construction of any circuit breaker requires the separation of contacts in an insulating medium. Under normal operating conditions these two contacts remain in closed position. For interrupting the circuit (both in normal or abnormal conditions), the moving contact moves away from the fixed contact and the flow of current is interrupted. This results in the formation of an arc between the contacts as shown in Fig. 11.1. The insulating medium between the contacts extinguishes the arc. It also provides adequate insulation between the contacts after separation and from each contact to earth.

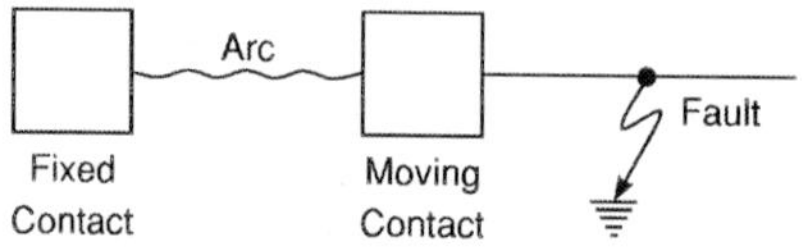

Fig. 11.1 Separation of contacts of circuit breaker

11.2 INITIATION OF ARC

When the moving contact of a circuit breaker begin to separate a large current passes through the gap. Since the contact resistance is very small, a small voltage appears across the gap. But the distance of separation is very small, a large voltage gradient is set up which is sufficient to cause the ionization of the particles between the contacts. This high voltage gradient remains only for few microseconds. But in this short period a large number of electrons are liberated from the cathode. This process of liberation of electrons is called field emission. These electrons during their travel to anode collide with the atoms and molecules of the gases to produce more electrons. As the contacts separate, the resistance between them increases and a lot of heat is produced due to ohmic losses. This heat is sufficient to ionize the air or vaporize and ionize the oil between the contacts. The ionized air or vapour acts as a conductor and the current remains uninterrupted across the low resistance arc produced. The voltage drop across the contacts is quite small and is just sufficient to maintain the arc. The resistance of an arc depends upon the number of particles ionized, the cross section of the arc and the length of the arc.

The voltage across the contacts during the arcing period is known as the ***arc voltage***.

11.3 ARC INTERRUPTION

There are two methods of arc interruption

1. High-Resistance Interruption
2. Low-Resistance or Current Zero Interruption

The basic difference between these two methods is the point of cut off of the current. In high resistance interruption, the current is cut off at any part of the current wave form, but the current is interrupted only at current zero in the low resistance interruption.

11.3.1 High Resistance Interruption

In this method of arc interruption, the resistance of the arc is increased to such a high value that it forces the current to a very low value insufficient to maintain the arc. The arc resistance may be increased due to deionizing by cooling, lengthening, constraining, and splitting the arc. This method is not suitable for large current interruption. This due to the fact that at the point of current interruption the energy associated with its magnetic field appears in the form of electrostatic energy producing a high voltage across the contacts of the circuit breaker. If this voltage is more than withstand capacity of the dielectric between the contacts the arc will restrike. Also, to tackle the sudden release of large amount of energy during the arcing period, the breaker requires good mechanical withstand capability. This method can be employed for both ac and dc low-power circuit breakers.

11.3.2 Low Resistance or Current Zero Interruption

In case of alternating current supply, the current wave passes through a zero point twice the supply frequency. Thus for a 50 Hz supply current crosses zero 100 times per second. This property of ac circuit is utilized for interruption purposes. The current in not interrupted at any point other than zero current instant and is not allowed to rise again after this zero occurs. If the current is interrupted at a point other than zero, high transient voltages will appear across the contacts of the circuit breaker.

11.4 ARC EXTINCTION PHENOMENON

The arc extinction phenomenon is explained by the following two theories:

1. Energy balance theory (Cassie's theory)
2. Recovery rate theory (Slepain's theory)

11.4.1 Energy Balance Theory

If the current is interrupted, heat is generated between the contacts. The heat generated varies with time depending upon separation of contacts. When the contacts are just opened at current zero, the restriking voltage is zero. Thus the heat generated is zero and when the contacts are opened fully and the arc is fully extinguished. The resistance between the contacts is almost infinite because the gap is fully deionized, thus heat generated is zero at this instant. Between these two instants, the

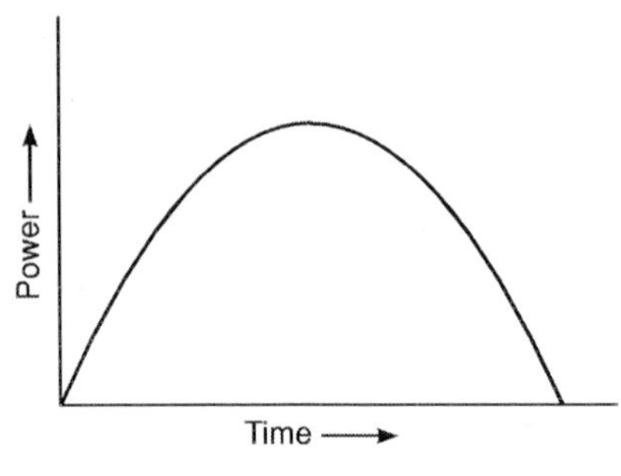

Fig. 11.2 Energy balance theory

power first increases, reaches a maximum value and then gradually decreases to zero as shown in Fig. 11.2.

According to Energy balance theory if the rate of heat generation is lower than the rate at which the heat is dissipated, the arc will be extinguished otherwise it will remain for another half cycle.

11.4.2 Recovery Rate Theory

The arc is due to the ionization of the gap between the contacts. This ionized medium consists of electrons and ions. To extinguish the arc, these electrons and ions must be removed from the gap immediately after the current reaches a natural zero. This is done either by chemically recombining these electrons and ions into neutral molecules or by flushing them away by inserting fresh insulation at a rate faster than the rate of ionization. The rate at which the gap recovers its dielectric strength is compared with the rate at which transient restriking voltage rises across the gap as shown in Fig. 11.3. If the rate of build up of dielectric between the contacts is faster than the rate at which recovery voltage rises the arc is extinguished, otherwise it will further restrike in next half cycle.

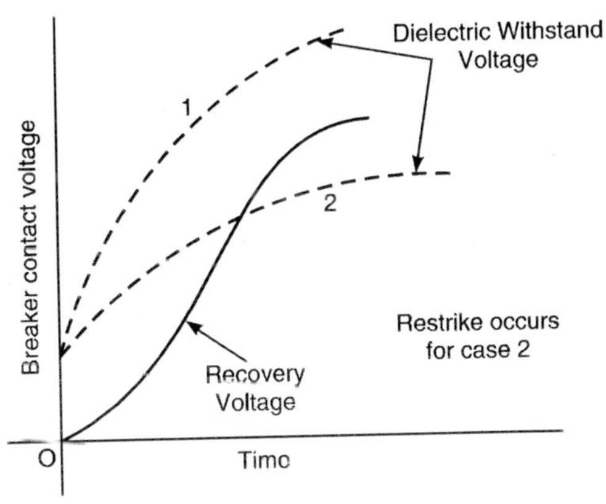

Fig. 11.3 Dielectric recovery in circuit breakers

11.5 RESTRIKING VOLTAGE AND RECOVERY VOLTAGE

The voltage across the circuit breaker contacts during the arcing period is called the arc voltage. This voltage becomes the system voltage when the arc is the extinguished. The arc is extinguished at the instant of current zero. After the arc has been extinguished, the voltage across the breaker contacts does not become the system voltage instantaneously, but it oscillates and there is a transient condition. At the instant of arc extinction, the resulting transient voltage which appears across the breaker contacts is quite high, and it will try to restrike the arc. Hence this voltage is known as ***restriking voltage***.

The power frequency rms voltage which appears across the circuit breaker contacts immediately after these oscillations die out and the arc finally extinguished is called the ***recovery voltage***.

Figure 11.4 shows the restriking and recovery voltages.

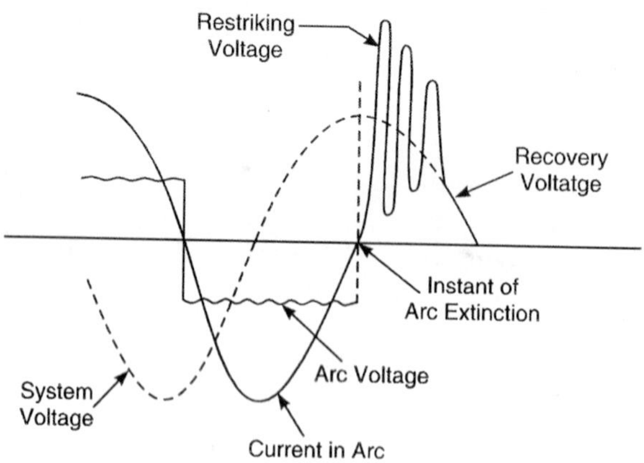

Fig. 11.4 Restriking and recovery voltage

11.6 EXPRESSIONS FOR RESTRIKING VOLTAGE AND RRRV

Transient conditions are produced when a system undergoes a change from one steady state to another steady state condition. It is due to the energy stored in the system inductance and capacitance. The amplitude of the transient wave depends upon the instant at which the circuit is disturbed. Restrike type switching transients occur during the opening operations. Figure 11.5 shows a power system and its equivalent circuit. L and C are the series inductance and the shunt capacitance up to the circuit breaker location respectively. In case of a fault, the short circuit current flows through L and the contacts of the circuit breaker. At power frequency the capacitance reactance is negligible but becomes significant at high frequency oscillations initiated by a sudden opening of the breaker.

Resistance is so small that it can be neglected. When the circuit breaker contacts are opened and the arc is extinguished, the current is diverted through the capacitance C. Now the inductance and capacitance of the line form a series oscillatory circuit. The voltage across the capacitor rises and oscillates at a natural frequency given by

$$f = \frac{1}{2\pi}\frac{1}{\sqrt{LC}}$$

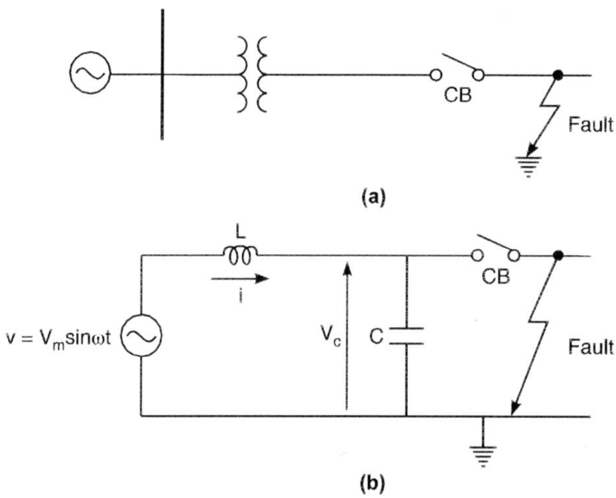

Fig. 11.5 (a) Fault on a feeder near circuit breaker (b) Equivalent electrical circuit

In this condition the voltage across the contacts of the circuit breaker is equal to the voltage across the capacitance. An expression of transient voltage can be found in terms of line inductance, capacitance, natural frequency of oscillation of the capacitor voltage and the system voltage.

Let the generator voltage be

$$V = V_m \sin \omega t \qquad ...(11.6.1)$$

Let V be the system voltage at the instant of arc interruption. As the transient oscillation is a fast phenomenon, V can be treated as a constant for a short duration.

By KVL in Fig. 11.5

$$v_L + v_c = V \qquad ...(11.6.2)$$

$$v_L = L\frac{di}{dt}$$

$$i = \frac{dq}{dt} = \frac{d(Cv_c)}{dt} = C\frac{dv_c}{dt} \quad ...(11.6.3)$$

$$v_L = L \cdot \frac{d}{dt}\left(C\frac{dv_c}{dt}\right)$$

$$v_L = LC\frac{d^2v_c}{dt^2} \quad ...(11.6.4)$$

∴ Eq. (11.6.2) becomes

$$LC\frac{d^2v_c}{dt^2} + v_c = V \quad ...(11.6.5)$$

Taking the Laplace transform of Eq. (11.6.5) we get

$$LCs^2V_c(s) + V_c(s) = \frac{V}{s}$$

$$V_c(s) = \frac{V}{s[LCs^2 + 1]}$$

$$V_c(s) = \frac{V}{LCs\left(s^2 + \frac{1}{LC}\right)} \quad ...(11.6.6)$$

Since $\omega_n = \frac{1}{\sqrt{LC}}, \quad \frac{1}{LC} = \omega_n^2$

$$V_c(s) = \frac{\omega_n^2 V}{s(s^2 + \omega_n^2)}$$

$$V_c(s) = V\left[\frac{1}{s} - \frac{s}{s^2 + \omega_n^2}\right] \quad ...(11.6.7)$$

Taking inverse Laplace of Eq. (11.6.7) we get

$$v_c(t) = V[1 - \cos \omega_n t] \quad ...(11.6.8)$$

or

$$v_c(t) = V\left[1 - \cos\frac{1}{\sqrt{LC}}t\right] \quad ...(11.6.9)$$

= restriking voltage

Arc restriking is mainly dependent upon

- maximum value of v
- maximum value of $v_c(t)$
- maximum value of rate of rise of $v_c(t)$.

From Eq. (11.6.9) the maximum value of restriking voltage occurs at

$$\omega_n t = \pi$$

or
$$t = \frac{\pi}{\omega_n} \qquad ...(11.6.10)$$

$$V_{Cmax} = V[1-(-1)] = 2V$$

$\therefore$ Maximum value of $v_c(t)$

$$V_{Cmax} = 2 \times \text{peak value of system voltage}$$
$$= 2V_m$$

The oscillatory transient voltage has a frequency of $\frac{1}{2\pi\sqrt{LC}}$ Hertz.

The rate of rise of restriking voltage

$$RRRV = \frac{dv_c}{dt} \qquad ...(11.6.11)$$

$$= \frac{d}{dt} V\,[1 - \cos \omega_n t]$$

$$RRRV = \omega_n\, V \sin \omega_n t \qquad ...(11.6.12)$$

The maximum value of $RRRV$ at instantaneous value V

$$= \omega_n V \qquad ...(11.6.13)$$

If the arc is interrupted at the instant of maximum system voltage, the maximum value of $RRRV$

$$(RRRV)_{max} = \omega_n\, V_m \qquad ...(11.6.14)$$

Example. 11.1 A 132 kV 50 Hz alternator is connected to a circuit breaker. The inductive reactance upto the circuit breaker is 4.5 Ω per phase. The distributed capacitance upto circuit breaker between phase and neutral is 0.01 μF. Calculate the following:

(a) Frequency of restriking voltage transient.

(b) The maximum value of restriking voltage across the breaker contacts.

(c) The maximum value of RRRV.

(d) Time at which the maximum value of transient recovery voltage occurs.

Solution :

$$X_L = 2\pi f L$$
$$4.5 = 2\pi \times 50 \times L$$

$$L = \frac{4.5}{2\pi \times 50} = 0.0143\text{H}$$

$$C = 0.01\ \mu\text{F} = 0.01 \times 10^{-6}\ \text{F}$$

(a) Frequency of transient oscillation

$$f_n = \frac{1}{2\pi\sqrt{LC}}$$

$$= \frac{1}{2\pi\sqrt{0.0143 \times 0.01 \times 10^{-6}}}$$

$$= 13.309 \times 10^3\ \text{Hz} = 13.309\ \text{KHz}$$

(b) Maximum value of restriking voltage

$$= 2 \times \text{maximum system voltage per phase}$$

$$= 2 \times \frac{132}{\sqrt{3}} \times \sqrt{2} = 215.56\ \text{kV}$$

(c) Maximum value of RRRV

$$= \omega_n \times (\text{maximum system voltage per phase})$$

$$= 2\pi f_n \times \left(\frac{132}{\sqrt{3}} \times 10^3 \times \sqrt{2}\right)$$

$$= 2\pi \times 13.309 \times 10^3 \times \frac{132}{\sqrt{3}} \times 10^3 \times \sqrt{2}$$

$$= 9012.937 \times 10^6\ \text{V/s}$$

(d) The time at which the maximum value of transient recovery voltage is given by

$$t = \frac{\pi}{\omega_n}$$

$$= \pi\sqrt{LC}$$

$$= \pi\sqrt{0.0143 \times 0.01 \times 10^{-6}}$$

$$= 0.03757 \times 10^{-3}\ \text{s}$$

$$= 37.57 \times 10^{-6}\ \text{s} = 37.57\ \mu\text{s}$$

11.7 INTERRUPTION OF CAPACITIVE CURRENT

The interruption of capacitive current produces high voltage transients across the gap of the circuit breaker. This occurs when an unloaded long transmission line or a capacitor bank is switched off. Consider the circuit shown in Fig. 11.6 (a).

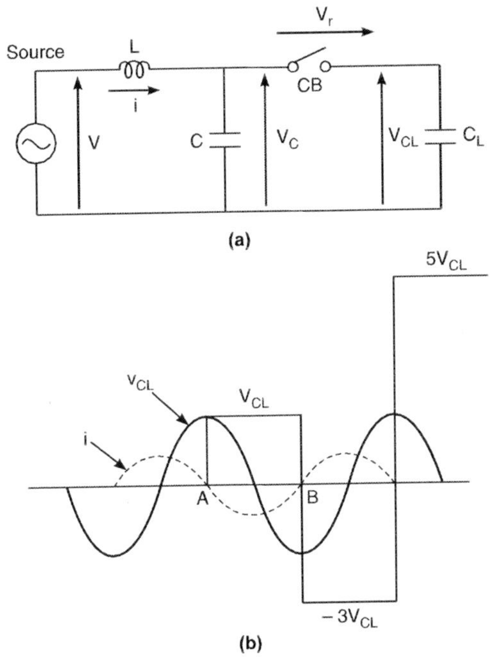

Fig. 11.6 Interruption of capacitive current

Let C = stray capacitance of the circuit breaker

C_L = line capacitance

Normally C_L is much larger than C. When capacitive current is interrupted suppose that

V_C = voltage across C

V_{CL} = voltage across C_L

V_r = voltage across the circuit breaker gap.

We know that the voltage across the capacitor cannot change instantaneously. The current supplied to a capacitor is usually small and the circuit breaker can interrupt such current at first current zero.

Fig 11.6 (b) shows transient voltage across the gap of the circuit breaker when capacitive current is interrupted.

The capacitive current lead the voltage by 90°. If the current is interrupted at A (the first current zero instant), the capacitor C_L is charged at the maximum value of the system voltage. It remains charged at this value.

After half cycle, that is, at instant B, the recovery voltage of approximate magnitude of V_{Cmax} across the circuit breaker and the total voltage across the circuit breaker is the sum of two voltages, that is,

$$V_{T\max} = V_{CL\max} + V_{r\max}$$

where $V_{T\max}$ = maximum voltage across the circuit breaker

V_{CLmax} = maximum voltage across the capacitor

$V_{r\max}$ = maximum value of power frequency recovery voltage

Thus, at instant B, that is half cycle from A, the voltage across the gap is twice the maximum value of V_C. Therefore a restrike is possible at this moment. If the arc restrikes, the voltage across the gap falls from $2\,V_{C\max}$ to zero. If a restrike occurs, the *LC* circuit will oscillate with a high frequency.

At this instant the capacitance C_L remains charged at -3 times the peak value of normal voltage that is $-3V_{CL\max}$. This sequence of events repeats once again and the capacitor voltage rises to $5V_{CL\max}$. The process continues indefinitely increasing the voltage by successive increments of $2V_{CL\max}$. It is only limited by leakage and corona losses or by breakdown of system insulation or capacitor puncture. In practice, the voltage across the gap is limited to four times the peak value of the system voltage.

11.8 CURRENT CHOPPING

When a circuit breaker interrupts low inductive current such as magnetizing current of a power transformer or current to a shunt reactor, current falls to zero before its natural zero. This phenomenon of interruption of current before natural zero is called ***current chopping***. It happens because a large deionizing force is produced by the circuit breaker. Current chopping occurs in fast acting air blast circuit breakers.

Consider the circuit shown in Fig. 11.7. Let the arc current be *I* when it is chopped down to zero. Let *L* be the inductance of the circuit and C be the stray capacitance. The energy stored in the inductance is $\frac{1}{2}Li^2$. The sudden change in current cannot make stored energy stored in the inductance zero instantaneously. This stored energy is transferred to the capacitance at the moment of current interruption.

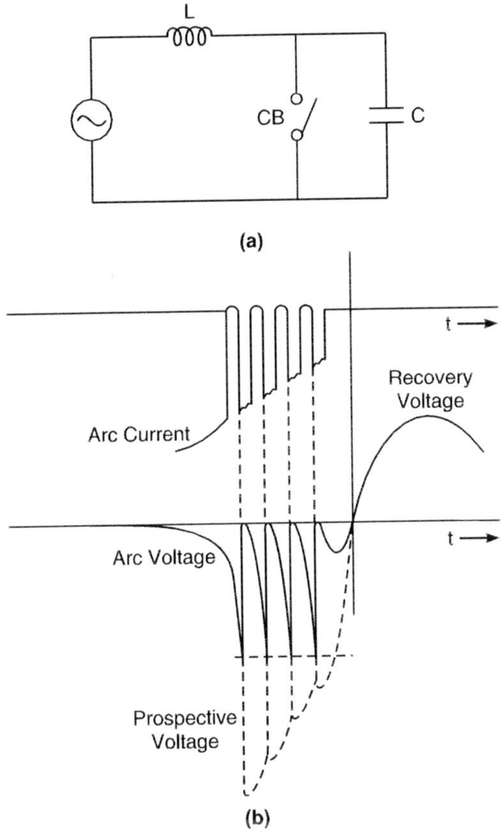

Fig. 11.7 Current chopping (a) circuit diagram, (b) chopping action

The energy transferred to C

$$= \frac{1}{2} Cv^2$$

where,

v = voltage developed across the breaker contacts

As the two energies are equal

$$\frac{1}{2} Li^2 = \frac{1}{2} Cv^2$$

or

$$v = i\sqrt{\frac{L}{C}}$$

The voltage v developed across the contacts of the circuit breaker due to transfer of energy is called the ***prospective voltage***. The prospective voltage may be extremely high as compared to the normal system voltage. If the value of v is more than the withstanding capacity of the gap between the contacts, the arc restrikes. The current is again chopped. This phenomenon continues till the value of v becomes less than the withstand capacity of the gap between the contacts. The chopping action is shown in Fig. 11.7 (b).

Resistance switching is used to overcome the effects of overvoltages due to current chopping.

Example 11.2 A three phase 50 Hz 20 MVA 110/11 kV transformer has no load current of 2.5 A and a stray capacitance of 0.05 μF. Calculate the maximum prospective voltage if the transformer is switched off at no load.

Solution : Maximum value of current to be interrupted

$$= \sqrt{2} \times 2.5$$
$$= 3.53 \text{ A.}$$

$$\text{Rated current} = \frac{20 \times 10^3}{\sqrt{3} \times 110}$$
$$= 105 \text{ A}$$

$$\text{No load current} = 2.5 \text{ A}$$

$$X_0 = \frac{110 \times 10^3}{\sqrt{3} \times 2.5} = 25.44 \times 10^3 \ \Omega$$

$$L_0 = \frac{25.44 \times 10^3}{314} = 81 \text{ H}$$

Prospective voltage at maximum value of no-load current

$$v = 3.53 \sqrt{\left(\frac{81}{0.05 \times 10^{-6}}\right)}$$
$$= 142.079 \text{ kV}$$

11.9 RESISTANCE SWITCHING

To damp out the transient oscillations in the LC circuit at the time of circuit breaker opening and to reduce the oscillation frequency, a resistance is connected across the contacts of the circuit breaker. This is known as ***resistance switching***. As the circuit breaker is opened a part of arc current flows through this resistance resulting in the decrease of the current through the contacts. Due to the decrease in the current through the contacts the restriking voltage and RRRV are reduced. This feature is used for problem free interruption of low inductive currents and capacitive currents. To make the process fast, sphere gap is used to switch in the resistance as shown in Fig 11.8.

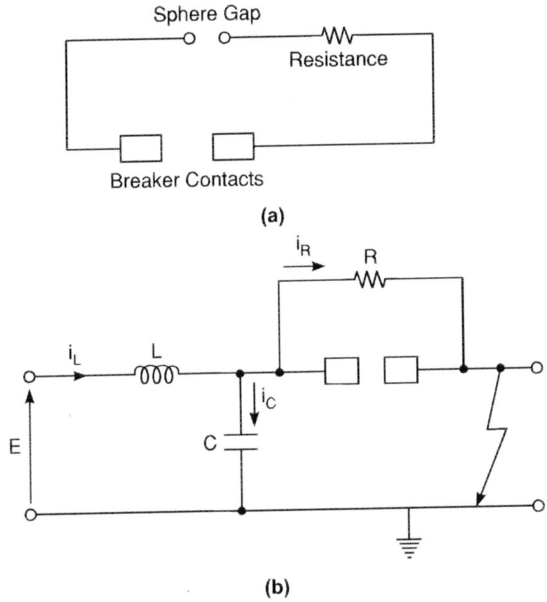

Fig. 11.8 Resistance switching

Figure 11.8 shows the equivalent circuit for analysis of resistance switching.

By *KCL*

$$i_L = i_c + i_R \quad ...(11.8.1)$$

By *KVL*

$$v_L + v_c = E \quad ...(11.8.2)$$

$$L\frac{di_L}{dt} + v_c = E \quad ...(11.8.3)$$

$$L\frac{d}{dt}(i_c + i_R) + v_c = E$$

$$L\frac{di_c}{dt} + L\frac{di_R}{dt} + v_c = E \quad ...(11.8.4)$$

But

$$i_c = \frac{dq}{dt} = \frac{d}{dt}(Cv_c)$$

$$i_R = \frac{v_c}{R}$$

$$\therefore \qquad \frac{di_R}{dt} = \frac{1}{R}\frac{dv_c}{dt}$$

substituting of these values in Eq. (11.8.4) gives

$$LC\frac{d^2 v_c}{dt^2} + \frac{L}{R}\frac{dv_c}{dt} + v_c = E \qquad ...(11.8.5)$$

Taking Laplace transform of Eq. (11.8.5) gives

$$LCs^2 V_c(s) + \frac{L}{R} sV_c(s) + V_c(s) = \frac{E}{s} \qquad ...(11.8.6)$$

$$LC\, V_c(s)\left[s^2 + \frac{1}{RC}s + \frac{1}{LC}\right] = \frac{E}{s}$$

or

$$V_c(s) = \frac{E}{sLC\left(s^2 + \frac{1}{RC}s + \frac{1}{LC}\right)} \qquad ...(11.8.7)$$

For no transient oscillations, all roots of Eq. (11.8.7) should be real. One root is zero, that is $s = 0$ which is real. For other roots to be real, the roots of the equation

$$s^2 + \frac{1}{RC}s + \frac{1}{LC} = 0 \qquad ...(11.8.8)$$

should be real. The following conditions should be satisfied.

$$\left(\frac{1}{RC}\right)^2 \geq \frac{4}{LC}$$

or

$$R^2 \leq \frac{L}{4C}$$

or

$$R \leq \frac{1}{2}\sqrt{\frac{L}{C}} \qquad ...(11.8.9)$$

Therefore, if the value of the resistance connected across the contacts of the circuit breaker is equal to or less than $\frac{1}{2}\sqrt{\frac{L}{C}}$ there will be no transient oscillations.

If

$$R > \frac{1}{2}\sqrt{\frac{L}{C}}$$

there will be no oscillations.

The value $R = \frac{1}{2}\sqrt{\frac{L}{C}}$ is called the ***critical resistance***.

Figure 11.9 shows the transient conditions for three different values of R . The frequency of damped oscillation is given by

$$f_n = \frac{1}{2\pi}\sqrt{\frac{1}{LC} - \frac{1}{4R^2C^2}}$$

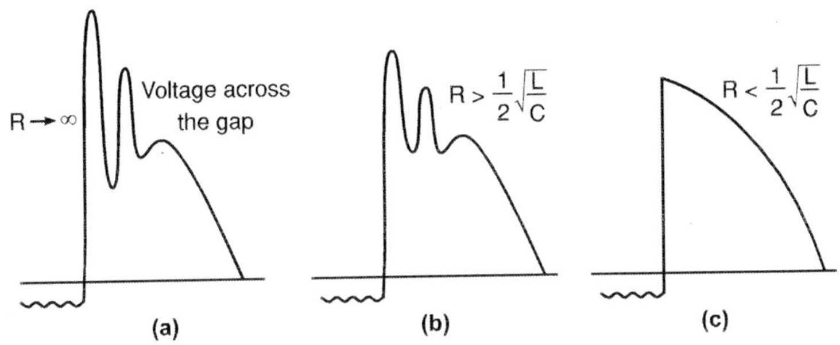

Fig. 11.9 Transient conditions for different values of *R*

The resistance of R performs one or more of the following functions:

1. It reduces the RRRV and thus reduces the burden on the circuit breaker.
2. It ensures the damping of the high frequency restriking voltage transients during switching of inductive or capacitive loads.
3. In a multibreak circuit breaker it helps in distributing the transient recovery voltage more uniformly across all the contact gaps.

Example. 11.3 In a 220 kV system, the inductive reactance and capacitance up to the location of circuit breaker are 10 Ω and 0.02 μ F respectively. A resistance of 800Ω is connected across the contacts of the circuit breaker. Determine the following:

(a) Natural frequency of oscillation,

(b) Damped frequency of oscillation.

(c) Critical value of resistance which will give no transient oscillation.

(d) The value of resistance which will give damped frequency of oscillation, one-fourth the natural frequency of oscillation.

Solution :

$$X_L = 2\pi f L$$

$$L = \frac{X_L}{2\pi f} = \frac{10}{2\pi \times 50}$$

$$= 0.03183\text{H}$$

(a) Natural frequency of oscillation

$$= \frac{1}{2\pi}\sqrt{\frac{1}{LC}}$$

$$= \frac{1}{2\pi}\sqrt{\frac{1}{0.03183 \times 0.02 \times 10^{-6}}}$$

$$= 6.037 \times 10^3 \text{ Hz.} = 6.307 \text{ kHz}$$

(b) Frequency of damped oscillation

$$= \frac{1}{2\pi}\sqrt{\frac{1}{LC} - \frac{1}{4R^2C^2}}$$

$$= \frac{1}{2\pi}\sqrt{\frac{1}{0.03183 \times 0.02 \times 10^{-6}} - \frac{1}{4 \times (800)^2 (0.02 \times 10^{-6})^2}}$$

$$= \frac{1}{2\pi}\sqrt{1570.8 \times 10^{-6} - 976.56 \times 10^{6}}$$

$$= 3.881 \times 10^3 \text{ Hz} = 3.881 \text{ kHz}$$

(c) Critical resistance

$$= \frac{1}{2}\sqrt{\frac{L}{C}}$$

$$= \frac{1}{2}\sqrt{\frac{0.03183}{0.02 \times 10^{-6}}} = 630.77\ \Omega$$

(d) Damped frequency of oscillation

$$= \frac{1}{4} \times \text{Natural frequency of oscillation}$$

$$= \frac{1}{4} \times 6.037 \text{ kHz}$$

$$= 1576 \text{ Hz}$$

Let R be the value of resistance which will have a damped frequency of oscillation equal to 1576 Hz

Then

$$1576 = \frac{1}{2\pi}\sqrt{\frac{1}{LC} - \frac{1}{4C^2R^2}}$$

$$=\frac{1}{2\pi}\sqrt{\frac{1}{0.03183\times 0.02\times 10^{-6}}-\frac{1}{4(0.02\times 10^{-6})^2\times R^2}}$$

Therefore, $R = 651.44\ \Omega$

Example 11.4 In a system of 132 kV the line to ground capacitance is 0.01 μF and the series inductance is 4H. Determine the voltage appearing across the pole of a circuit breaker when a magnetizing current of 5A in interrupted. Calculate the resistance to be connected across the contacts to eliminate the restriking voltage.

Solution : The voltage appearing across the pole of the circuit breaker

$$v = i\sqrt{\frac{L}{C}}$$

$$= 5\times\sqrt{\frac{4}{0.01\times 10^{-6}}}$$

$$= 100\times 10^3\ \text{V}$$

$$= 100\ \text{kV}$$

Critical resistance to be connected across the contacts to eliminate the restriking voltage.

$$\text{R} = \frac{1}{2}\sqrt{\frac{4}{0.01\times 10^{-6}}}$$

$$= 10\times 10^3\ \Omega$$

$$= 10\ \text{k}\Omega$$

11.10 INTERRUPTION OF SHORT LINE FAULT

A fault located a few kilometers from the circuit breaker is called a ***short-line fault*** or ***kilometric fault***. Short-line faults are characterized by high-frequency of restriking voltage of the order of 10 to 100kHz depending upon the length of the line and location of the fault. The voltage appearing across the breaker terminals after final current interruptions is the phasor sum of the voltage from the source and the high frequency voltage from line side. The peak value of the high frequency component is reached in a few microseconds. Hence the rate of rise of the transient recovery voltage is very high. This can result in arc restriking.

EXERCISES

1. Explain various duties performed by a circuit breaker. Describe various methods of arc extinction.
2. Discuss different methods of interrupting the arc current in circuit breakers. Explain two main theories of current zero interruption.
3. Explain the terms
   (a) Recovery voltage
   (b) Restriking voltage
   (c) Rate of rise of restriking voltage (RRRV)
4. Explain the significance of RRRV in operation of a circuit breaker. Also derive the relation for RRRV considering simple power system.
5. Discuss the problems associated with the interruption of
   (a) Low inductive current
   (b) Capacitive current
   (c) Fault current if fault is very near to the substation
   and suggest their remedies.
6. What is resistance switching ? Derive the expression for critical resistance in terms of system inductance and capacitance which gives no transient oscillation.
7. Explain the phenomenon of current chopping in a circuit breaker. What means are taken to reduce it ?

❑❑❑

12

CIRCUIT BREAKERS

12.1 CLASSIFICATION OF CIRCUIT BREAKERS

Circuit breakers are classified on the basis of the medium employed to extinguish the arc. They are

1. Air Circuit Breakers
2. Oil circuit Breakers
3. Sulphur Hexafluoride (SF_6) Circuit Breakers
4. Vacuum Circuit Breakers

12.2 AIR-BREAK CIRCUIT BREAKERS

Air-break circuit breakers use the principle of high resistance for arc interruption. As explained in the previous chapter the arc resistance can be increased by lengthening, splitting and cooling the arc. Arc runners, arc splitters, arc horns and magnetic blow-outs (provided by blow-out coils) are employed for this purpose. In air-break circuit breakers relatively high arc resistances are obtained near the current zeros. The power factor near the current zero is modified to reduce the restriking voltage.

Figure 12.1 shows a simple air-break circuit breaker. Two pairs of contacts per phase are used. The main contacts made of a low resistance material carry current under normal conditions. Another set of contacts usually made of tungsten (high resistance material) is used to carry the total current during the process of interruption.

When the contacts are opened, the main contacts separate first, followed by arcing contacts. This results in shifting of the main current to the arcing contacts causing an arc between the arcing contacts. The arc is now drawn into the arc splitters by thermal and electromagnetic effects resulting in quenching of the arc.

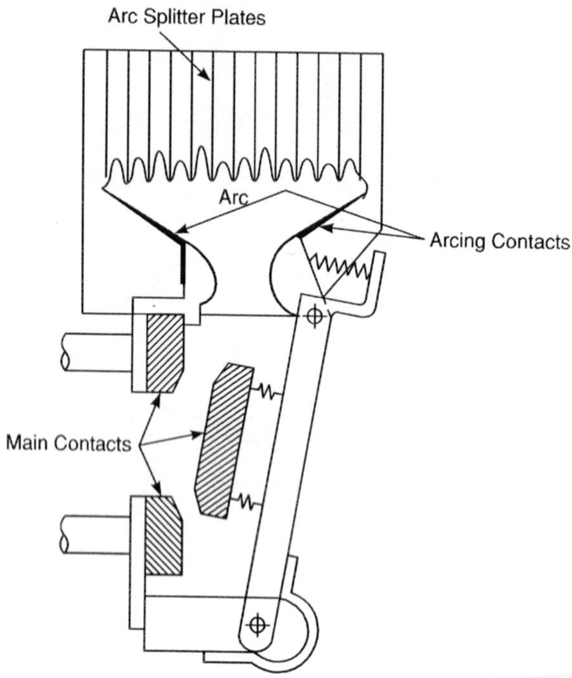

Fig. 12.1 Air break circuit breaker

Air break circuit breakers are used in low and medium voltage range applications. They are available in the voltage range of 0.4 to 11 kV and preferred over oil circuit breakers for places prone to fire hazards.

12.3 AIR BLAST CIRCUIT BREAKERS (ABCBs)

In air blast circuit breakers, compressed at pressures of 2-6 MN/m^2 is used to extinguish the arc. Air blast circuit breakers are classified as

1. Axial blast type
2. Cross blast type

12.3.1 Axial Blast Type Air Blast Circuit Breaker

In this type of breaker, a high pressure blast of air is longitudinally forced on the arc. Such an arrangement is shown in Fig. 12.2.

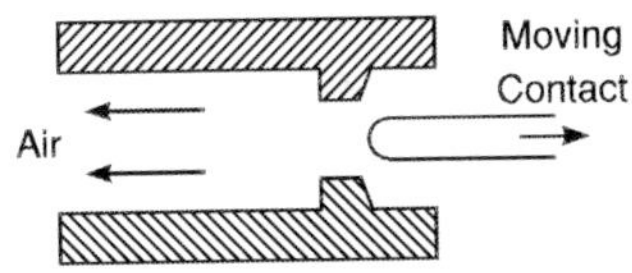

Fig. 12.2 Single axial blast principle

At current zero as the tripping command is initiated the air valve opens to force the compressed air at high speeds directed longitudinally towards the contacts. Due to this forced air the diameter of the arc is greatly reduced. The energy content of the arc is proportional to the square of the arc diameter and heat loss is proportional to the arc diameter (surface area of arc). Considerable heat loss takes place due to forced convection and the arc extinguishes as the diameter of the arc goes on reducing. Hollow contact arrangement is preferred for high power circuits.

Double blast or radial blast principle is shown in Fig. 12.3

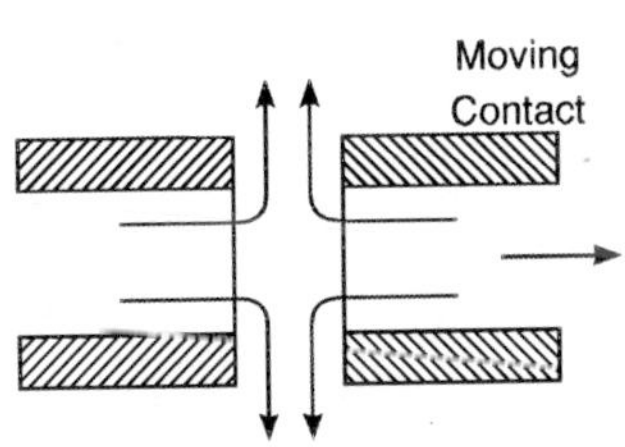

Fig. 12.3 Double or radial blast circuit principle

Single and double axial blast circuit breakers are used with EHV and UHV systems voltages upto 1200 kV.

Generally more than one break per phase is employed for voltages above 33 kV. Switched resistors are connected across the contacts to reduce the transient voltages. Capacitors are also connected across the contacts (interrupters) to equalize the voltage across the breaks, in case more than one break per phase is used.

12.3.2 Cross Blast Air Circuit Breakers

In a cross blast air circuit breaker (Fig. 12.4) the air blast is directed across the arc. The air blast forces the arc into the arc splitters. These plates cause lengthening and splitting of the arc. This process is similar to air-break circuit breakers where the arc is forced into the splitters with the help of electromagnetic and thermal effects. In a cross-blast design, resistance switching is not frequently employed. This is because in lengthening of the arc without undue restriction, sufficient resistance is present in the arc itself to control any restriking. Cross blast type breakers are generally not preferred for high current interruptions.

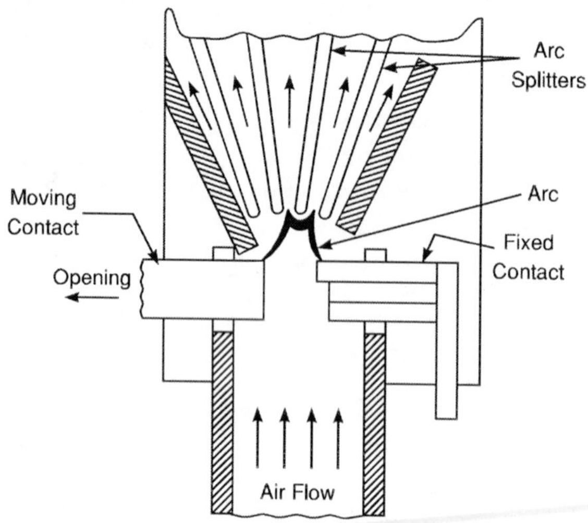

Fig. 12.4 Cross air blast circuit breaker

Advantages of air blast circuit breakers (ABCBs)

The advantages of ABCBs are:

- Suitable for reliable, frequent and high speed operations.
- Cheap and abundantly available interrupting medium.
- No deterioration of interrupting medium as fresh medium is used for every interruption.
- Short and consistent arcing duration, therefore, less burning of contacts and less maintenance.
- High speed reclosure possible.
- Fire hazard is eliminated.

Disadvantages of air blast circuit breakers (ABCBs)

The disadvantages of ABCBs are :

- Current chopping and restriking possibilities.
- Noisy operation if proper silencers are not provided.
- Air compressing plant required plus additional maintenance required.

12.4 OIL CIRCUIT BREAKERS

In an oil circuit breaker the contacts are separated in insulating oil (mineral oil). Mineral oil has better insulating and cooling properties as compared to air. When an arc is formed, the arc energy is absorbed in decomposition of oil. Oil decomposes into gases like hydrogen, acetylene etc. These gases cause a high pressure inside the arcing chamber, resulting in turbulent flow of gases and oil which in turn quenches the arc. The oil circuit breaker can be subdivided into

1. Bulk oil circuit breakers
2. Low oil or minimum oil circuit breakers

On the basis of arc extinction, oil circuit breakers are classified as

(a) Plain break oil circuit breaker

(b) Self-generated pressure oil circuit breaker

(c) Externally generated pressure or impulse type oil circuit breaker

Bulk oil circuit breakers generally use plain break mechanism. Self and externally generated pressure mechanisms are used for low or minimum oil circuit breakers.

12.4.1 Plain Break Oil Circuit Breaker

In a plain break oil circuit breaker, the fixed and moving contacts are enclosed in a strong, weathertight earthed and securely bolted tank filled with oil. An air cushion is provided between the oil surface and the tank cover to accommodate the oil displaced due to formation of gases due to the arc. It absorbs the mechanical shock due to upward movement of the oil.

As the contacts start to open in oil, the separation is very small in the beginning, a high voltage gradient between the contacts ionize the oil. Large volumes of gases are produced as the oil decomposes and expands as gas bubbles surrounding the arc region. It pushes away the oil away from the arc.

The arc quenching depends upon

- The elongation of the arc
- The head of oil above the arcing contacts.
- Turbulent motion of oil due to the gases formed.
- Speed of contact separation.

The speed of contact travel should be kept as high as possible. This is due to the fact that a certain distance of contact travel is required before the interruption is likely to occur. Sooner this condition is reached, smaller is the energy released which in turn imposes lesser mechanical shock to the breaker. For this purpose two breakers in series are provided which give rapid arc lengthening even without high speed contacts. Such an arrangement is called double break arrangement and is also useful for breaking low currents.

In plain break oil circuit breakers the oil performs the function of arc extinguishing as well as serves as an insulating medium between the live parts and earth. Thus, the plain break oil type circuit breakers are bulk oil circuit breakers.

12.4.2 Self-generated Pressure Oil Circuit Breakers

In these types of circuit breakers the contacts are made to open in a chamber made of an insulating material, known as explosion pot or pressure chamber or arc controlling extinction device (AED). The arc energy is utilized to generate a high pressure in the chamber.

The pressure developed inside the chamber depends upon the magnitude of the current. This creates a problem in designing of AEDs, because for low values of currents the pressure generated should be sufficiently enough to extinguish the arc and for high values of current the pressure must not be high enough to blast the pot. The advantages of using AEDs are

- The breaking capacity of the circuit breaker is increased.
- Arcing time is reduced.
- Compact assembly.

Various designs of explosion pots are developed and used to suit different requirements. Some important ones are as follows.

Plain explosion pot

Figure 12.5 shows a plain explosion pot. In case of a fault, the moving contact moves vertically out of the chamber through a tight fitting throat at the lower end of the pot. The high pressure causes a turbulent flow of gases into the arc, resulting in arc extinction. If the arc is not extinguished within the pot the ***axial movement*** of gases along the arc at the throat of the chamber causes arc extinction. Therefore sometimes plain explosion pot is also known as ***axial blast pot***.

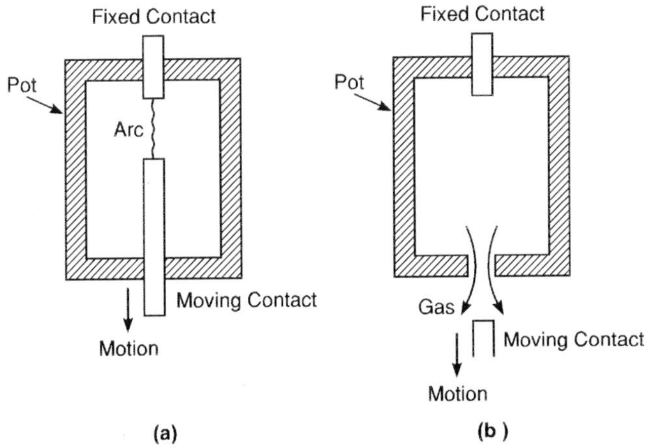

Fig. 12.5 Plain explosion pot

This type of pot construction is used for breaking medium level currents as high currents cause a pressure high enough to burst the pot and arcing time is increased in breaking low currents.

Cross-jet explosion pot

Figure 12.6 shows a cross jet explosion pot. This type of pot consists of metallic channels across the moving contact. These metallic channels are used as arc splitters and vents to the gases formed due to the arc. Since the high pressure gases are expelled across the arc through these vents, this type of pot is known as cross-jet explosion pot.

This type of pot construction is suitable for high current interruptions as the gases formed are regularly expelled through the vents as the moving contact moves away. It is to be noted that the moving contact does not leave the chamber as in case of axial blast pot.

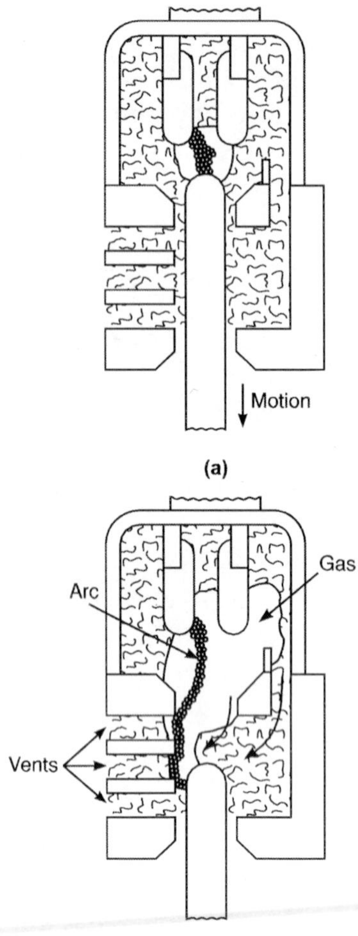

Fig. 12.6 Cross jet explosion pot showing four stages of explosion

Self compensated explosion pot

Figure 12.7 shows a self compensated explosion pot. It is a combination of a cross-jet and axial-blast explosion pot. This type of pot construction is suitable for low as well as high currents. The upper portion is similar to a cross-jet pot which works satisfactorily for cases of high current interruptions and lower portion of the chamber similar to axial-blast pot is useful to extinguish low level currents.

When the current to be interrupted is low, the rate of gas formation is low and builds up only up till the tip of the moving contact starts leaving the chamber and extinguishes the arc on axial-blast principle. For large currents the rate of gas formation is very high and a high pressure is developed very quickly and the pot works as a cross-jet pot.

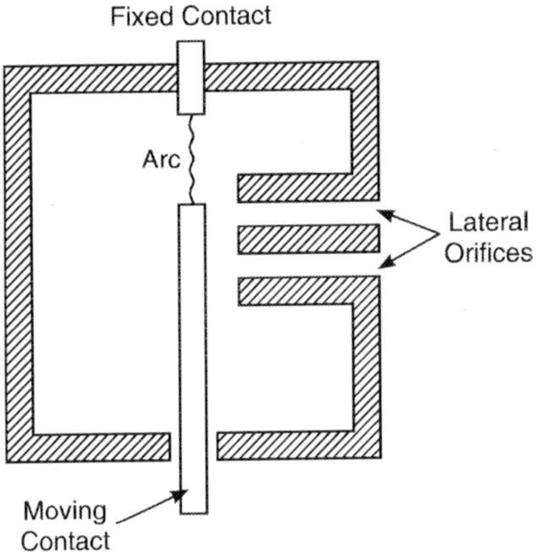

Fig. 12.7 Self compensated explosion pot

It is to be noted that the lateral vents are automatically closed as soon moving contact crosses these vents. So sufficient pressure is developed if a low current is interrupted on an axial-blast principle.

Since the pot adjusts itself according to the current level to be interrupted therefore it is known as ***self compensated explosion pot***. Self compensated pot designs are used in low content or minimum oil circuit breakers. In minimum oil circuit breakers (MOCB) the oil performs the function of arc extinction and some solid insulation serves as an insulating medium between live parts and earth resulting in a compact design. MOCBs are used for voltages upto 220kV.

12.4.3 Externally Generated Pressure or Impulsive Type Circuit Breakers

Some mechanical piston cylinder type arrangement is used to develop the required oil pressure to extinguish the arc. The piston is mechanically coupled to the moving contact and the oil pressure is created as soon as moving contact starts separating from the fixed contact.

12.4.4 Double Break Oil Circuit Breakers

A double break oil circuit breaker employs an intermediate contact between the fixed and the moving contact. As soon as the trip command is initiated the intermediate contact starts moving away with the moving contact, drawing an arc between the fixed and the intermediate contact. As soon as the intermediate contact stops, a second arc is initiated between the intermediate contact and the moving contact. The second arc is extinguished by the gases produced due to the first arc. Both axial blast and cross blast principles use double-break construction. This is shown in Fig. 12.8.

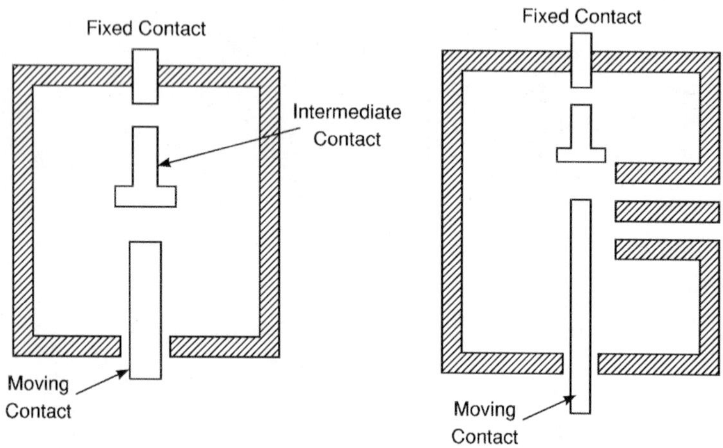

Fig. 12.8 Double break oil circuit breakers
(a) Axial blast pot (b) Cross-blast pot

Disadvantages of oil circuit breakers

- Oil and arc products are inflammable
- Dielectric strength of oil is reduced by carbonization during the arcing process.
- Oil is hygroscopic in nature. Moisture in oil deteriorates the dielectric properties of oil.

12.5 SULPHUR HEXAFLUORIDE (SF_6) CIRCUIT BREAKERS

12.5.1 Properties Of Sulphur Hexafluoride Gas (SF_6)

Sulphur hexafluoride (SF_6) is noninflammable, non-toxic, inert and chemically very stable gas. It has a good dielectric strength and is an excellent arc quenching medium.

SF_6 is an electronegative gas hence it can acquire electrons easily. When a free electron comes in contact with a neutral gas molecule the electron is absorbed and forms a negative ion. As the negative ion thus formed is very much heavier as compared to free electron, its mobility is very much reduced. Thus, it can no more take part in ionization process of the gas. Also the gas permits the removal of free electron from a neutral molecule or negative ion at relatively high temperatures. These properties of SF_6 make it a good dielectric. The dielectric strength of SF_6 gas is about 2.5 times that of air at atmospheric pressure. It is comparable to that of transformer oil at pressure of 3 atmospheres.

SF_6 decomposes into SF_4, SF_2, S_2, F_2, S and F at temperatures around 1000°C. SF_6 has got good recombination properties after dissociation and removal of source which energizes the arc.

The low dynamic time constant (about 1μsec i.e., 100 times shorter than air) gives an excellent quenching property. The gas also has an excellent heat transfer property and its thermal time constant is 1000 times shorter than air.

Sulphur hexafluoride circuit breakers (SF_6 CBs) can withstand severe RRRVs as the dynamic time constant of the gas is very small (i.e., for the same value of limiting voltage the natural frequency of the mains may be much higher).

Two major disadvantages of SF_6 gas is that in presence of moisture, the decomposition products can corrode some parts of the circuit breaker which are in contact with the gas and SF_6 gas condensates at low temperatures and the liquidification is pressure dependent. To overcome these drawbacks a mixture of soda lime (NaOH + CaO) and activated alumina are placed in arcing chamber to absorb any moisture or byproducts and controlled heaters are installed when working in low ambient temperatures. SF_6 breakers are used for voltages upto 765 kV.

Advantages of SF6 *Circuit Breakers*

1. The gas and its decomposition products are chemically stable. Hence there is no danger of fire or explosion.
2. The smaller size of conductors and clearances lead to small breaker size (due to high dielectric strength of the gas).
3. They have a good overload margin due to excellent heat transfer property of the gas.

4. Very low maintenance required. Contacts and other components do not corrode due to inertness of the gas and short time arc.
5. Low gas velocity and pressures are required and employed prevent current chopping and capacitive currents can be interrupted without restriking.
6. Operation is silent as there is no high pressure gas exhaust to atmosphere like ABCB.

12.5.2 Types of SF_6 Circuit Breakers

SF_6 breakers can be of two types

(i) Single pressure SF_6 circuit breakers

(ii) Double pressure SF_6 circuit breakers.

Single pressure SF_6 circuit breakers can be sub-divided into two categories

(a) Puffer-piston system based circuit breakers

(b) Self blast system based circuit breakers

Puffer-piston type SF_6 circuit breakers

Figure 12.9 shows a puffer-piston type circuit breaker in a closed position. The gas is sealed inside the body of the breaker, at a pressure of about 3 atmospheres. When the contacts open, a gas blast at high pressure, to extinguish the arc is obtained by a puffer mechanism. Puffer mechanism is a piston-cylinder arrangement in which the cylinder moves at a high speed into the cylinder thereby forcing the gas through nozzles at high speeds into the low pressure region. The cylinder is mechanically coupled to the operating mechanism. To quench the arc the puffer-piston arrangement blows the arc axially.

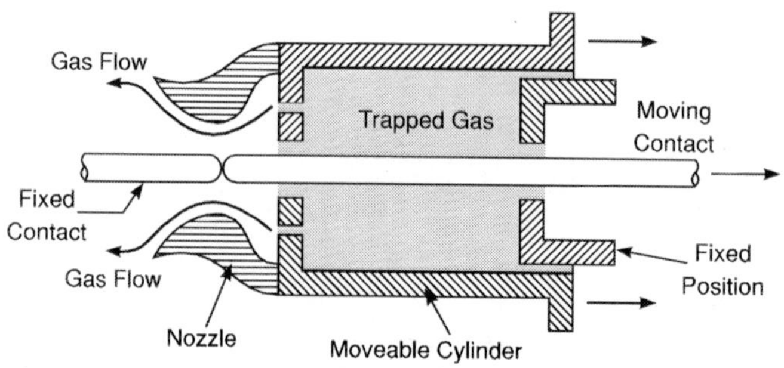

Fig. 12.9 Puffer-piston type SF_6 circuit breakers

Some design employing puffer-piston mechanisms provide a combination of axial and cross blast together are also developed.

Self blast type SF_6 circuit breaker

Separate contacts are used for interruption purposes and carrying the load current. Using separate contacts provides continuous current carrying capability to the breaker. During an opening operation the current is transferred to an insulated coil wound around the load contacts. The magnetic field produced by the current causes the arc to revolve. The heat produced by the arc creates an SF_6 gas overpressure in the chamber. This causes a turbulent axial flow of the SF_6 gas which results in quenching of the arc at current zero.

In double pressure SF_6 breakers the functions of insulation and interruption is performed by SF_6 in different chambers at different pressures. Presently double pressure SF_6 breakers are not used because of their design complications and a large number of auxiliaries needed.

12.6 VACUUM CIRCUIT BREAKERS

Vacuum is a space condition in which the pressure of the space is kept below the atmospheric pressure. Under normal conditions atmospheric pressure is taken equal to 760 mm of mercury. Pressures of the order of 10^{-5} mm is employed as arc interrupting medium in vacuum circuit breakers. Under this pressure the mean free path (the distance an electron travels without any collision) is very small and is of the order of few metres. The mean free path of electrons in gas (arc quenching medium used in other types of circuit breakers) is small and the electron ionization process multiplies to form an electron avalanche. A movement of contacts by a few millimeters in high vacuum cannot cause any possible electron collision. Thus the formation of arc due to the formation of an electron avalanche is not possible in high vacuum.

In a vacuum arc the electrons and ions do not come from the medium in which the arc is initiated. The separation of current carrying contacts causes the metal vapour to be released from the contact surface liberating electrons and ions for maintaining the arc.

The breakdown strength is dependent upon

- Gap length
- Condition of the electrode surface

Highly polished electrodes are used in vacuum circuit breakers. Even then when an arc is drawn, the arcing does not take place on the entire contact surface but is initiated only at a few micro projected points known as ***hot spots***. Different contact geometries are employed to overcome the above problem.

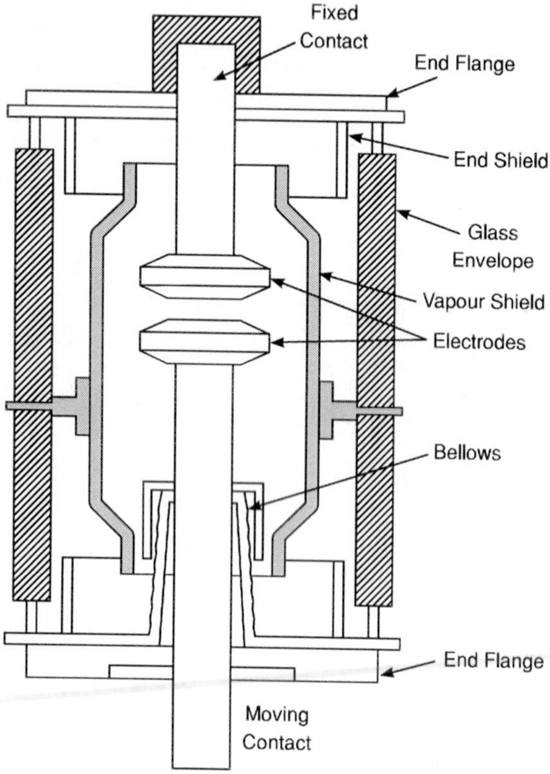

Fig. 12.10 Vacuum circuit breaker

After the first current zero the metal vapour disperses and the vacuum regains its dielectric strength extinguishing the arc in a half cycle.

Figure 12.10 shows a schematic diagram of a vacuum circuit breaker. It consists of a fixed and a moving contact enclosed in a vapour shield to prevent the metal vapour from condensing on the outer glass or bellow surface. Metal bellows made of stainless steel are used to move the moving contact without any loss of vacuum.

Vacuum circuit breakers are used for where severe RRRV conditions are involved such as in capacitor bank switching, reactor switching, transformer switching etc. Vacuum circuit breakers are suitable for voltages of upto 33 kV and are compact, cheap and reliable for frequent switching operations.

12.7 OPERATING MECHANISM

Proper opening and closing of a circuit breaker largely depends upon the type and design of the operating mechanism used. The main functions of an operating mechanism are:

- Rapid and smooth closure for all currents from rated to making capacity.
- To hold the contacts in closed position until a trip command is received in any circuit condition.
- To open the contacts without any delay as soon as the trip command is received.
- To perform the auto reclosure cycle.

The opening operation should be a fast phenomenon. The fault clearing process is dependent on it. The opening system must ensure movement of contacts at a proper speed to achieve the desired opening characteristics and must also damp the movement of the contacts at the end of the travel.

Friction, stiction, electromagnetic forces and forces due to inertia and quenching medium oppose the opening operation.

Closing of contacts during normal conditions is a simple phenomenon. If the circuit breaker is made to close contacts under short circuit conditions large electromagnetic forces repel the contacts. Also friction, inertia of moving system and opposing forces due to the medium also hinder the closing operation. The operating mechanism should have enough potential to overcome these forces and provide a speedy closure of the contacts. Above all for reliable operation, the operating mechanism should be independent on the main supply.

The operating mechanisms used are

(1) Spring

(2) Solenoid

(3) Compressed air

The characteristics of these operating mechanisms are shown in Fig 12.11.

Springs are very good for opening system as they produce a large force in the beginning and gradually decrease as shown in Fig. 12.11 This property is disadvantageous in closing operations. For small capacity breakers the spring recharging during closure is done by the hand. For large ones it is done with the help of as motor or by closing mechanism.

Solenoids are very good for closing as the force of attraction increases as the distance between the contacts decreases. They are not suitable for opening operations as their action is very slow.

The force remains constant along the contact travel in case of compressed air operating mechanisms. Hence, compressed air operating systems are suitable for opening as well as closing purposes. Air blast circuit breakers generally employ compressed air operating mechanisms.

Sometimes a combination of compressed air and spring or solenoid and spring mechanisms is used for closing and opening operations respectively.

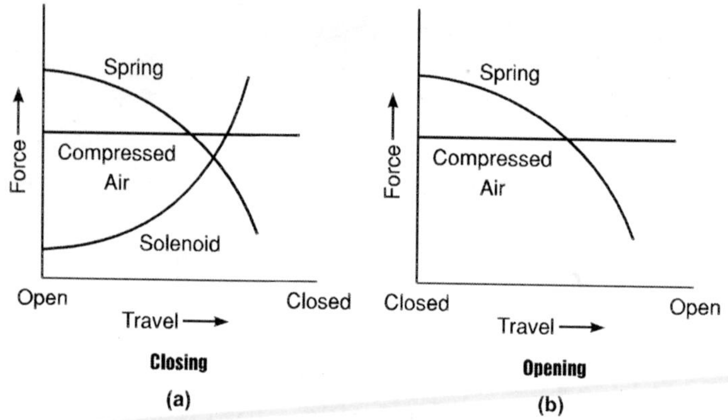

Fig. 12.11 Characteristics of operating mechanisms

12.8 HVDC CIRCUIT BREAKING

HVDC transmission lines have a number of advantages as compared to ac transmission lines. Such lines are used for point to point transmission of large amount of electrical power which is also called two terminal HVDC systems. Faults on these systems are cleared by controlling the firing angle of the thyristors employed in converters. Also the switching operations are performed on the ac side with ac circuit breakers. HVDC circuit breakers are neither essential nor either used in two terminal HVDC systems. On systems with more than two terminals, faults of transient nature are cleared by adjusting the switching of the thyristors used in converters. The permanent faults are cleared by a combination of thyristor control, fault locators and isolating switches. Hence **HVDC circuit breakers are required for parallel HVDC lines, HVDC lines with taps or a closed loop HVDC system.**

A number of HVDC switching methods are developed and used. Two of them are.

(a) Almost all HVAC circuit breakers employ zero current switching. This feature is not available in dc. If a large HVDC is interrupted abruptly, a high transient voltage appears across the contacts of the breaker. HVDC can be interrupted by creating artificial zeros of current through the contacts of the switch with the help of some external circuits. These circuits produce an oscillatory current across the breaker contacts. The crest currents of the oscillation must be greater than the dc to be interrupted.

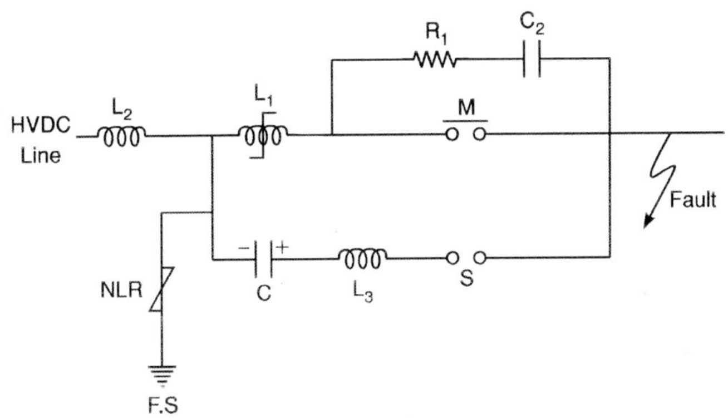

Fig. 12.12 Circuit diagram for direct-current interruption

Figure 12.12 shows the schematic diagram of HVDC circuit breaker with artificial current zero principle. The main circuit breaker M is either a SF_6 or a vacuum circuit breaker. The saturable core reactor L_2 and the series circuit R_1C_2 are employed to shape current before interruption and the transient voltage after interruption. L_1 is the line inductance. C and L_3 are used to produce artificial current zero.

At the instant the main circuit breaker M starts opening, the vacuum gap S is triggered, causing the precharged capacitor C to discharge into the closed circuit. It sends a current in a direction opposite to the short circuit current. As this current increases, it will force the short circuit current to zero with a few oscillations. The arc is interrupted at a suitable current zero.

(b) HVDC can also be interrupted by the scheme shown in Fig.12.13.

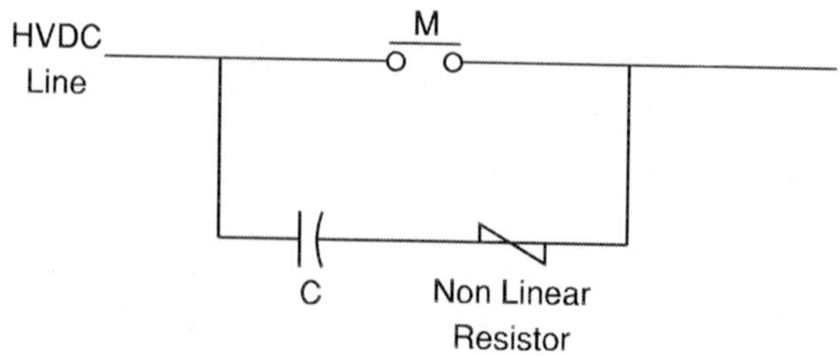

Fig. 12.13 HVDC breaker with nonlinear resistor in parallel

In case of normal operating conditions the current flows through the main circuit breaker M. The capacitor is initially uncharged. In case of a fault, the main contacts open and the current is diverted to the capacitor, thereby limiting the current to be interrupted by the main circuit breaker to a very small value. The nonlinear resistor absorbs and dissipates the energy without increasing the voltage across the contacts of the breaker.

12.9 RATINGS OF CIRCUIT BREAKERS

A circuit breaker must work satisfactorily under normal conditions and wide range of abnormal conditions. In addition to rated voltage, current, frequency and duty cycle, a circuit breaker has three major duties to perform under short circuit. They are

1. It must be capable of opening the contacts on occurrence of faults to clear the fault (*Breaking capacity*).
2. It must be capable of closing the contacts onto a fault. (*Making capacity*).
3. It must be capable of carrying the fault current for a short time while another circuit breaker in series is clearing the fault. (*Short-time capacity*).

12.9.1 Breaking capacity

Breaking capacity is expressed in terms of the highest current a circuit breaker is capable of breaking under specified conditions of recovery voltage. The breaking capacity of a breaker is the product of breaking current and the recovery voltage (generally considered equal to the rated voltage). For a three phase circuit breaker it is given by

Breaking capacity $= \sqrt{3} \times$ rated current $\times$ rated voltage

It is generally expressed in MVA. It is of two types

(i) Asymmetrical breaking capacity

(ii) Symmetrical breaking capacity

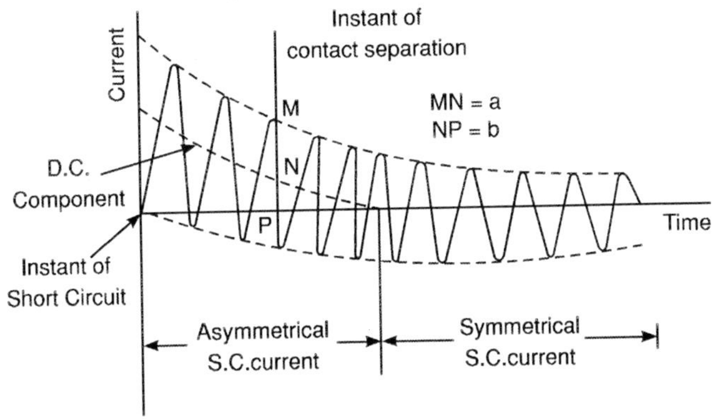

Fig. 12.14 Short circuit current waveforms

Consider a short-circuit current wave of one phase of a system. It consists of an ac component and a dc component. The dc component dies out after a few cycles depending upon the damping involved. Then the current becomes symmetrical.

The rms value of the fault current (comprising of both ac and dc components) that a circuit breaker can break under specified conditions of recovery voltage is known as ***asymmetrical breaking capacity***. From Fig.12.14 it can be expressed as

$$I_{asymmetrical} = \sqrt{\left(\frac{a}{\sqrt{2}}\right)^2 + b^2}$$

where

$\frac{a}{\sqrt{2}}$ = rms value of the ac component at the instant of contact separation

b = value of dc component at the instant of contact separation.

The rms value of ac component of the fault current that a circuit breaker is capable of breaking under specified conditions of recovery voltage is known as ***symmetrical breaking capacity.***

12.9.2 Making Capacity

The rated making current is defined as the peak value of the current (including both ac and dc components) at the instant the circuit breaker can be closed onto a short circuit. It is expressed as

$$I_{making} = \sqrt{2} \times 1.8 \times \text{symmetrical breaking current}$$

$$= 2.55 \times \text{symmetrical breaking current}$$

RMS value is converted into the peak value by multiplying it by $\sqrt{2}$ and a factor of 1.8 takes into account the dc component.

12.10 TESTING OF CIRCUIT BREAKERS

The two important tests performed on a circuit breaker are:

1. Routine tests
2. Type tests

12.10.1 Routine Tests

Routine tests are conducted on each piece of circuit breaker to confirm its proper functioning. The routine tests confirm the quality of the circuit breaker and reveal the defects if any in components due to manufacturing of circuit breaker. These tests are classified as

(a) Mechanical operation tests

(b) Millivolt drop test and measurement of resistance

(c) Power frequency voltage tests

(d) Testing of auxiliary and control circuits

12.10.2 Type Tests

Type tests are performed to prove the rated characteristics of a circuit breaker. These tests are carried out on a few randomly selected pieces from a whole lot or a batch, made to the same specifications or ratings. These tests include

(a) Dielectric tests

(b) Short circuit tests

(c) Thermal tests

(d) Mechanical tests.

Dielectric Tests

These tests are conducted for testing the insulators of the circuit breaker. These include:

(A) Power frequency tests

These tests are of two types

(i) One minute dry withstand test.

(ii) One minute wet withstand test

(B) Impulse voltage withstand test

In power frequency tests the test voltage is applied for a period of one minute between

- Phases with breaker closed
- Phase and earth with breaker open
- Across the poles with breaker open

Water for artificial rain is sprayed on external insulation to verify the wet withstand tests in the specified conditions. Wet type tests are carried out for outdoor switchgears only.

Generally impulse voltage withstand tests are performed on outdoor switchgears only. The high frequency switching surges also occur on indoor switchgears, but their effect is greatly suppressed in cable systems used for indoor switchgears. To consider the effect of overvoltages caused by lightning, a test voltage wave of standard 1/50 μs (according to Indian standards) is applied, the peak value of which is specified according to the rated voltage of the breaker. This test voltage waves of both polarities is applied across

- each pole and earth with other phases earthed and breaker closed
- all terminals on one side of the circuit breaker and with other side terminals solidly earthed with open breaker.

The breaker must not flashover for ten such applications.

Thermal Tests

Rated current is passed through the closed circuit breaker, through all the phases for a long period to achieve steady state conditions. Temperature rise of various conducting, insulating and structural parts are taken at regular intervals by means of thermocouples. The maximum temperature rise must be in permissible limits.

In addition, measurement of contact resistance is also carried out at these temperatures.

Mechanical Tests

Repeated opening and closing of the breaker is done to ensure that the breaker must open and close at the required speed without any mechanical failure. The number of such operations may go up to a few thousands to check the possible causes of failure and the probabilities of their occurrence. No adjustments in the breaker are done during the testing period.

Short circuit Tests

Short circuit tests come under type tests. They are conducted to prove the ratings of a circuit breaker and aiding research and development through experimental investigation of problems of circuit breaking. These tests include:

(i) Breaking current tests
(ii) Making current tests
(iii) Short time current tests
(iv) Operating duty tests
(v) Basic short circuit test duties
(vi) Short line fault test
(vii) Breaking of capacitive and small inductive currents

There are two types of testing methods to carry out these tests.

1. Direct Method

These are of two types

a. Field testing
b. Laboratory testing

2. Indirect Method

Direct tests are done on low and medium capacity circuit breakers and indirect tests are done on large capacity breakers.

To carry out these tests specially built short-circuit testing plants are required.

Direct testing plants are of three types.

1. Field type
2. Laboratory type
3. Composite type.

In field type testing station, the circuit breaker to be tested is directly connected to a large power system. A major advantage of this method of testing is that it is economical particularly for HV circuit breakers.

The drawback of this method is lack of flexibility in testing conditions. They are:

- The tests cannot be repeatedly carried out for research and development work without disturbing the power system.
- The power available depends upon the installed capacity loading conditions and location of the station.
- Recovery voltage and RRRV can not be controlled as may be desired.

In a laboratory type testing station, specially designed generators, installed in the laboratory, provide the required power for testing. Its advantages are:

- Desirable test conditions i.e., current, voltage, restriking voltage RRRV and power factor can be controlled accurately.
- The test can be repeated again and again as required for research and development work.
- Indirect testing methods can be used for large capacity circuit breakers.

Its drawbacks are:

- Installation cost of station is very high.
- Limited power is available for testing of circuit breakers.

A composite testing station is a combination of field type testing station and laboratory type testing station.

Basic Short Circuit Testing Station

A simplified diagram of a testing station is shown in Fig. 12.15. The function of various components is described below.

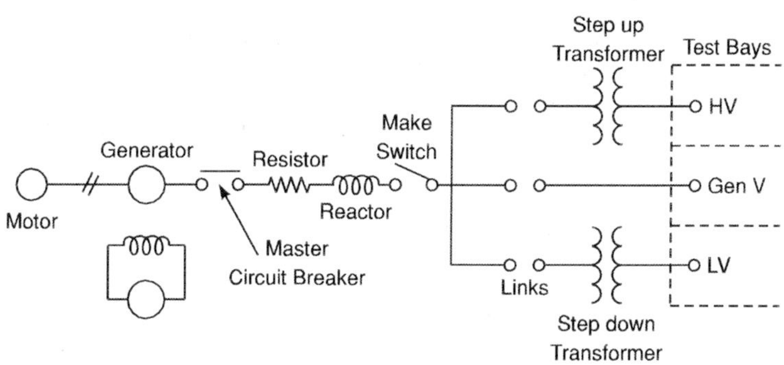

Fig. 12.15 Single line diagram for a short-circuit test plant

Short circuit generator

Full short circuit power is supplied by this generator. The generator is driven by a three phase induction motor. A very rigid bracing of conductors and coils ends is provided in view of high electromagnetic forces developed during short circuit conditions. The short-circuit generator is specially designed to have low reactance to give maximum short-circuit output. A flywheel is also provided on the generator shaft to supply kinetic energy during short-circuits and helps in speed regulation.

Impulse excitation

A short circuit current has a demagnetizing effect on the generator main flux due to its lagging power factor thereby reducing the main flux. Consequently the generator's emf is reduced. At the time of short circuit the field current is increased to about 10 times its normal value to counter the demagnetizing effect of armature reaction.

Short circuit transformer

This transformer is designed to have low leakage reactance. Due to its intermittent use, cooling requirements are reduced to minimum. It is designed to withstand repeated short circuits. Three phase transformers are used for step down purposes. A bank of single phase transformers is employed for voltages higher than the generated voltage.

Master Circuit breaker

It provides protection against failure of test breaker. It is set to operate at a predetermined time after the initiation of the start circuit and isolates the circuit breaker under test from the supply source after every test. This breaker must have an interrupting capacity in excess of the short circuit MVA of the supply source.

Resistor and reactor

These are used to control the power factor and the short circuit current. They also control the transient recovery voltage (TRV).

Capacitor

Capacitors control RRRV and can also be used for the purpose of voltage measurement.

Make switch

It controls the instant of short circuit on the voltage wave and is closed after the master circuit breaker and test circuit breaker.

All the components are adjusted so as to obtain the desired values of current, voltage, RRRV and power factor as discussed earlier. The short circuit tests are performed as follows

Breaking Capacity Test

After the master circuit breaker and the circuit breaker to be tested are closed, the short circuit current is interrupted by opening the test breaker at the desired moment by closing the make switch.

The following measurements are taken from the oscillogram connected across the test breaker.

(i) Breaking capacity (symmetrical and asymmetrical)
(ii) Recovery voltage and RRRV
(iii) Amplitude factor and natural frequency.

Making Capacity Test

First the master circuit breaker and the make switch are closed and short circuit is applied by closing the breaker under test.

Making current can be calculated from the oscillographic record as explained earlier.

Short Time Current Test

The rated short-circuit current is passed through test circuit breaker for a short duration (1 sec or 3 sec as specified) and the oscillographic

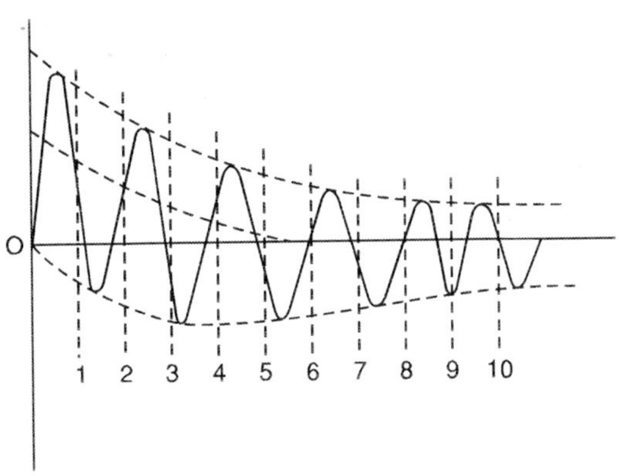

Fig. 12.16

record of the current is used to measure the current. This current should not cause any damage (insulation or mechanical) to the test breaker. The breaker should be capable of carrying the full load current continuously without excessive temperature rise after the test.

To calculated the equivalent rms short time current, the recorded oscillograph of the short time current is divided into 10 equal parts marked by 0,1,2,....,10 as shown in Fig. 12.16. The asymmetrical rms values of the current at these intervals are marked as $I_0, I_1, I_2, \ldots, I_{10}$. By Simpson's formula the equivalent rms value of the short time current can be calculated with sufficient accuracy and is given by

$$I=\sqrt{\frac{1}{3}\left[I_0^2+4(I_1^2+I_3^2+I_5^2+I_7^2+I_9^2)+2(I_2^2+I_4^2+I_6^2+I_8^2+I_{10}^2)\right]}$$

Duty Cycle Test

The duty cycle tests performed are as follows

(i) B – t – B – t – B duty test is performed at 10%, 30%, and 60% of the rated symmetrical breaking capacity.

(ii) B – t – MB – t – MB duty test is performed at not less than 100% of the rated symmetrical breaking capacity and rated making capacity.

This test duty can be tested separately as

(a) M – t – M for make test

and

(b) B – t – B – t – B for break test

(iii) B – t – B – t – B duty test is performed at not less than 100% of rated asymmetrical breaking capacity.

Here B, M and MB denote breaking operation, making operation and making operation followed by breaking operation respectively without any intentional time delay. 't' denotes the time in minutes between successive operations of an operating duty.

Several re-takes may be carried out to confirm the ratings.

Indirect Testing

Direct tests are not suitable for large capacity high voltage circuit breakers as the testing station required should be of a very large capacity which is highly uneconomical. Also increasing the short circuit capacity of the testing station is impractical. Therefore indirect methods of testing are used for testing high capacity high voltage circuit breakers. The important indirect methods of testing include:

1. Unit testing
2. Synthetic testing

Unit Testing

High voltage circuit breakers consist of several arc interrupter units in series per pole. These interrupters operate simultaneously. Generally the breaking capacity in MVA is shared equally by these units. Each unit is tested separately to determine the capacity of the breaker. This is known as ***unit testing*** or ***element testing***.

Synthetic testing

Synthetic testing provides a practical and economical solution for the testing of high voltage circuit breakers of large capacities. A synthetic testing circuit generates the electrical stresses those impressed on a circuit breaker during the interruption of actual fault current without actually employing the corresponding short circuit.

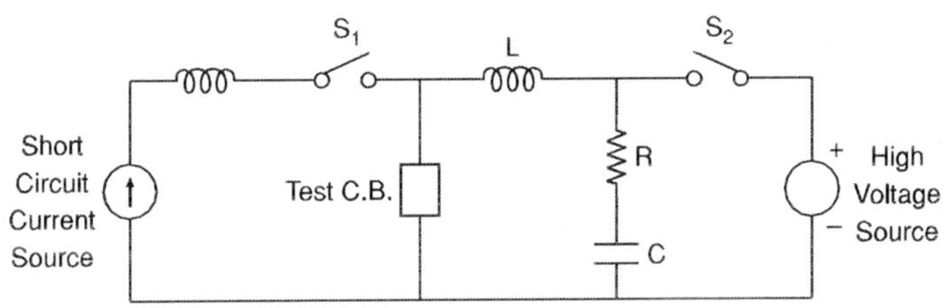

Fig. 12.17 Basic synthetic testing circuit

Figure 12.17 shows a basic synthetic testing circuit for testing circuit breakers. The current (low voltage) source provides the short circuit current. The voltage (low current) source provides the restriking and recovery voltages. L, C and R are used to tune the desired tests conditions. The switch S_1 is closed to supply the short circuit current. Near the final current zero switch S_2 is closed. The type of synthetic testing depends on whether S_2 is closed before or after current zero. In current injection synthetic testing S_2 is closed before current zero and S_2 is closed after current zero in voltage injection type.

Current injection can be divided in two types

1. Parallel current injection
2. Series current injection

These are classified on the basis of whether the voltage circuit is connected in parallel or series with the test breaker respectively.

Parallel current injection method is widely accepted as it is capable of providing RRRV and recovery voltage required by various standards accurately.

A parallel current injection circuit is shown in Fig. 12.18. The master CB, auxiliary CB and test CB are closed. The make switch is switched on to initiate the short circuit. The short circuit current is controlled by an inductor L_i.

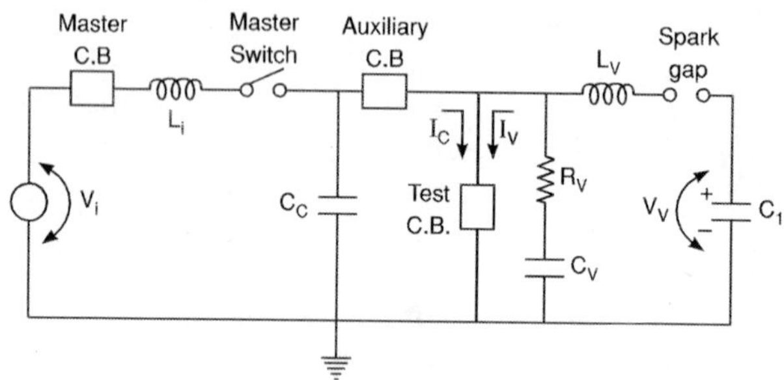

Fig. 12.18 Parallel current injection method

The inductance L_V, capacitor C_V, resistance R_V, a spark gap and a pre-charged capacitor C_1 forms the voltage circuit.

At a predetermined time the auxiliary and test breakers start opening and the arc is initiated at their contacts as shown in Fig. 12.19.

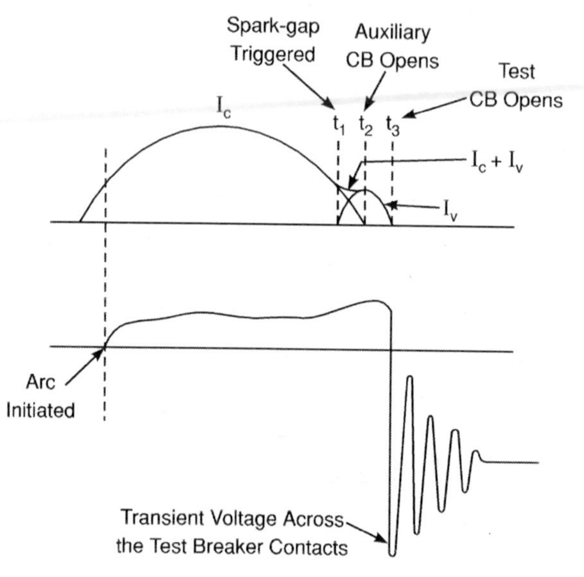

Fig. 12.19 Current and voltage waveforms of synthetic testing

The spark gap is triggered slightly before the short circuit current I_C reaches its natural zero. It forces the current I_V into the test current breaker. L_V and C_V are adjusted to control RRRV and recovery voltage. The current $(I_C + I_V)$ flows during the $t_1 - t_2$. The auxiliary breaker opens at t_2 and I_C becomes equal to zero. At t_3 the test circuit breaker opens and $I_V = 0$.

The frequency of the voltage circuit is given by

$$f = \frac{1}{2} \pi \sqrt{L_V C_V}$$

and the transient voltage of this frequency appears across the test breaker. The short-circuit current and the amplitude and frequency of the transient voltage can be varied over a wide range.

EXERCISES

1. What are the differences between air blast circuit breakers and air break circuit breakers ?
2. Explain the construction and working of an air blast circuit breaker.
3. What is multi break circuit breaker? How voltage distribution is equalized across the breaks?
4. What is the function of an explosion pot in an oil circuit breaker. Differentiate between different types of explosion pots based on their construction, operating principle and application.
5. How minimum oil circuit breaker is different from bulk oil circuit breaker? Gives its advantages and disadvantages.
6. Explain the construction and working of a puffer type SF_6 circuit breaker. Also enumerate the properties which makes SF_6 a good dielectric and arc quenching medium.
7. Describe the behaviour of electric arc in vacuum. With the help of a neat diagram explain the construction of a vacuum circuit breaker.
8. What are the different types of operating mechanisms used for opening and closing of the contacts of a circuit breaker? Discuss their merits and demerits.
9. Discuss the problems encountered in HVDC circuit breaking. Suggest remedies for them.
10. Explain how making capacity, breaking capacity and short time capacity are determined.

11. What are various types of rating of a circuit breaker. Explain the terms
    (a) Symmetrical and Asymmetrical breaking capacity
    (b) Making capacity
    (c) Short time current capacity
12. What are the different tests carried out on a circuit breaker ? Differentiate between type tests and routine tests.
13. What is the difference between direct testing and indirect testing? Also describe the various procedure of indirect testing.
14. With a neat sketch describe the layout of a simple short circuit plant. Also give details of equipments used.
15. Explain with a neat diagram the principle of parallel current injection for synthetic testing.

❑❑❑

INDEX

A

AC motor protection 108

Admittance relay 75

Advantages of neutral grounding 127

Air-blast circuit breakers (ABCBs) 196

Air-break circuit breakers 195

Amplitude comparators

- averaging type 32
- phase splitting type 34
- rectifier bridge type 31

Amplitude comparators 31

Angle impedance relay 79

Application of overcurrent relays 61

Application of directional overcurrent relays 61

Arc interruption 177

- current zero interruption 178
- high-resistance interruption 177
- low-resistance interruption 178

Arc voltage 177

Arc suppression coil 132

Arcing grounds 127, 156

Autoreclosing 118

- single shot 120
- multishot 120
- single-phase 121
- single-pole 121
- three-phase 121

B

Backup protection **8**

Basic Impulse Insulation level (BIIL) **171**

Basic insulation level (BIL) **171**

Basic requirements of protective relaying **6**

Biased differential protection **75**

Busbar protection **116**

C

Choice of grounding **135**

Classification of protective schemes **10**

Classification of relays **10**

Counterpoises **161**

Critical flashover voltage **172**

Cross-jet explosion pot **201**

Curent transformers (CT) **12**

Current chopping **186**

Current grading **53**

Current tap setting **48**

Coordination delay time (CDT) **55**

D

Differential protection of busbars **116**

Differential protection **81**

Direct strokes **159**

Directional relay **57**

Directional control **62**

Directional overcurrent **63**

Distance relays **68**

Duality between phase and amplitude comparators **41**

E

Effect of cable on surge **150**

Electromagnetic relays

- balanced beam relay **19**
- induction cup type **24**
- induction disc type **23**
- plunger type **19**
- polarized moving coil **21**
- polarized moving iron **22**
- shaded pole type **22**
- wattmetric type **23**

Electrothermal relays **27**

Energy balance theory **178**

Expulsion type surge diverter **165**

External overvoltages **158**

F

Faults **2**

- effects **4**
- nature and causes **2**
- open conductor **4**
- series **3**
- shunt **3**
- symmetrical **3**
- unsymmetrical **3**

Fibre-optic links **115**

Flat line **144**

Free neutral system **123**

G

Gapless surge diverter **168**
General relay equation **57**
Generator protection
    percentage differential **96**
    stator earth fault **99**
    stator interturn fault **100**
    rotor earth fault **101**
    loss of excitation **102**
    unbalanced currents **104**
    overheating **105**
    prime mover failure **106**
Grounding transformer **134**
Ground fault neutralizer **132**

H

HVDC circuit breaking **210**

I

Impedance relay **69**
Impulse ratio **172**
Induced strokes **159**
Infinite line **144**
Instrument transformers **11**
    advantages **11**
    main functions **11**
Insulation coordination **173**
Internal overvoltages **155**
Induction type directional overcurrent relay **59**

Isokeraunic level **158**

Isolated neutral system **123**

K

Kilometric fault **193**

L

Linear couplers **116**

Loss of excitation protection **102**

M

Mechanism of lightning discharge **144**

Merz Price principle **82**

Metal-oxide surge diverter **168**

Mho relay **75**

Microwave pilot protection **115**

Modified impedance relay **77**

Modes of coupling **114**

O

Objectives of power system protection **5**

Offset mho relay **78**

Ohm relay **79**

Ohmic setting of distance relays **67**

Oil circuit breakers **199**

Overcurrent relay **45**

Overhead earth wire **161**

Overhead ground wire **161**

Overfluxing protection of transformers **93**

P

Percentage differential protection **83**
of generators **96**
of transformers **83**
Peterson-coil grounding **130**
Phase angle error of CT **13**
Phase comparators
coincidence type **35**
integrating type **38**
phase splitting type **36**
pulse signal type **40**
rectifier bridge type **37**
vector product type **35**
Phase comparison carrier current protection **113**
Pickup current **48**
Pilot relaying **110**
Pilot-wire protection **111**
Pickup setting **48**
Plain explosion pot **201**
Plug setting multiplier (PSM) **49**
Potential transformer (PT) **15**
Power-line carrier protection **112**
Primary protection **8**
Prime mover failure **106**
Protective angle **162**
Protective zones **7**
Protection of parallel feeders **61**

R

R–X diagram **67**
Rating of circuit breakers **212**

Ratio error of CT **12**

Reach of relays **65**

- overreach **65**
- underreach **65**

Reactance grounding **130**

Reactance relay **72**

Recovery rate theory **179**

Recovery voltage **180**

Refraction coefficient **142**

Relay comparators

- amplitude comparators **31**
- phase comparators **35**

Relay coordination **51**

Resistance grounding **129**

Resistance switching **188**

Restriking voltage **180**

Reverse power relay **58**

Reflection coefficient **141**

Reflection factor **141**

Rod gap **163**

Rotor earth fault protection **101**

RRRV **180**

S

Selectivity of relays **6**

Selection of a communication channel **115**

Shielding angle **162**

Solid grounding **128**

Stator earth fault protection **99**

Stator interturn fault protection **100**

Static relays **29**
Sulphur hexafluoride (SF_6) circuit breaker **204**
Surge **139**
Surge diverters
    expulsion type **165**
    valve type **166**
    metal oxide **168**
    zinc oxide **168**
Switchgear **1**
System neutral ground **123**

T

Testing of circuit breakers **234**
Three-zone protection **65**
Time current grading **54**
Time grading **52**
Time multiplier setting (TMS) **50**
Time-current characteristics **46**
    extremely inverse **47**
    definite-time **46**
    instantaneous time **46**
    inverse definite minimum time **46**
    inverse time **46**
    very inverse **47**
Transformer protection **86**
    Buchholz relay **91**
    desensitizing scheme **90**
    harmonic current restraint **88**
    magnetizing inrush current **87**
Transmission coefficient **142**

U

Unit generator-transformer protection **107**

V

Vacuum circuit breaker **207**
Valve type surge diverter **166**
Voltage transformer (VT) **15**
Volts/hertz protection **94**

W

Waveshapes of stroke currents **159**

Z

Zinc oxide (ZnO) diverter **168**